Supplying Installations

Electrical Installation Series – Advanced Course

M. Doughton

Edited by Chris Cox

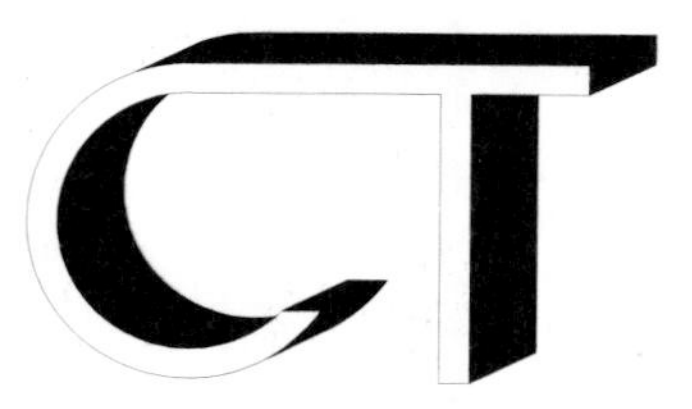

Australia · Canada · Mexico · Singapore · Spain · United Kingdom · United States

Supplying Installations

For more information, contact Thomson Learning, Berkshire House, 168–173 High Holborn, London, WC1V 7AA or visit us on the World Wide Web at: http://www.thomsonlearning.co.uk

---

British Library Cataloguing-in-Publication Data
A catalogue record for this book is available from the British Library

---

ISBN 1-86152-725-X

First published 2001 by Thomson Learning

Printed in Croatia by Zrinski d.d.

About this book

"Supplying Installations" is one of a series of books published by Thomson Learning related to Electrical Installation Work. The series may be used to form part of a recognised course, particularly City and Guilds 2360 Electrical Installation Course C, or individual books can be used to update knowledge within particular subject areas. A complete list of titles in the series is given below.

Electrical Installation Series

Foundation Course

Starting Work
Procedures
Basic Science and Electronics

Supplementary title:
Practical Requirements and Exercises

Intermediate Course

The Importance of Quality
Stage 1 Design
Intermediate Science and Theory

Supplementary title:
Practical Tasks and Revision Exercises

Advanced Course

Advanced Science
Stage 2 Design
Electrical Machines
Lighting Systems
Supplying Installations

Acknowledgements

The author and publishers gratefully acknowledge the following illustration source:

Bussman Division, Cooper (U.K.) Ltd. (p.28)

Every effort has been made to trace all copyright holders but if any have been inadvertently overlooked, the publishers will be pleased to make the necessary arrangements at the first opportunity.

Study guide

This studybook has been written to enable you to study either in a classroom or in an open or distance learning situation. To ensure that you gain the maximum benefit from the material you will find prompts all the way through that are designed to keep you involved with the subject. If you are studying by yourself the following points may help you.

- ☞ Work out when, and for how long, you can study each week. Complete the table below and from this produce a programme so that you will know approximately when you should complete each chapter. Your tutor may be able to help you with this. It may be necessary to reassess this timetable from time to time according to your situation.
- ☞ Try not to take on too much studying at a time. Limit yourself to between 1 hour and 2 hours and finish with a Try this or the short answer questions at the end of the chapter. When you resume your study go over this same piece of work before you start a new topic.
- ☞ You will find answers to the questions at the back of the book but before you look at the answers check that you have read and understood the question and written the answer you intended.
- ☞ An end test is included so that you can assess your progress.
- ☞ Try this activities are included and you may need to ask colleagues at work or your tutor at college questions about practical aspects of the subject. These are all important and will aid your understanding of the subject.
- ☞ It will be helpful to have available for reference a current copy of BS 7671, Guidance Note 1 Selection and Erection and the Electricity Supply Regulations 1988. At the time of writing BS 7671 incorporates Amendment No.1, 1994 (AMD8536), Amendment No. 2, 1997 (AMD 9781) and Amendment No. 3 (AMD 10983) 2000.
- ☞ Your safety is of paramount importance. You are expected to adhere at all times to current regulations, recommendations and guidelines for health and safety.

Study times					
	a.m. from	to	p.m. from	to	Total
Monday					
Tuesday					
Wednesday					
Thursday					
Friday					
Saturday					
Sunday					

Programme	Date to be achieved by
Chapter 1	
Chapter 2	
Chapter 3	
Chapter 4	
Chapter 5	
Chapter 6	
End test	

Contents

1

Statutory and Non-statutory Regulations

At this point at the beginning of each chapter you will be asked to complete a revision exercise based on the previous chapter. These should be successfully completed before you progress to the next topic.

To start you off we will use this opportunity to remember some facts that you should be aware of.

The majority of electrical installations are supplied via the national supply network, provided by Public Electricity Suppliers. Some installations are equipped with their own generation equipment, either as first line supplies or as standby generators, for use in the event of a supply failure.

The provision of the public electricity supply network is governed by statutory regulation which details such information as the tolerance acceptable on voltage and frequency. The regulation also places an obligation on the supplier to maintain supplies to the consumer and details actions required when these are not achieved.

Supply companies are involved in the provision and maintenance of the supply equipment up to the point where the consumer receives the supply into their installation. For small consumers this is normally at the nominal voltages of 400/230 V. Large consumers may receive their supply at considerably higher voltages.

On completion of this chapter you should be able to:

- demonstrate knowledge of the Electricity Supply Regulations 1988 and BS 7671 relating to supplies of electricity
- state the need to refer to appropriate British Standards, Statutory Regulations and Codes of Practice when carrying out design
- recognise the requirements of the Supply Companies to comply with the Electricity Supply Regulations 1988
- state the need for distribution systems under the control of the consumer to comply with the Electricity Supply Regulations 1988 and The Electricity at Work Regulations 1989

Electricity Supply Regulations

In this chapter we shall be considering the requirements of the statutory and non-statutory regulations in relation to the supplies of electrical energy. In particular we shall look at the Electricity Supply Regulations 1988, including all amendments, and BS 7671 Requirements for Electrical Installations (IEE Wiring Regulations 16th Edition), with particular reference to supply. The Electricity Supply Regulations (Amendment) (No.2) Regulations 1994 make some changes to the 1988 Regulations. The amendment introduces the change to Regulation 30 which determines the nominal voltage between phase and neutral conductors at the supply terminals. It further sets out the permitted variations on declared frequency and voltages, the latter banded as below 132 kV or 132 kV and above.

It would be an advantage to have a copy of the Electricity Supply Regulations, including amendments, for reference during the study of this chapter. Whilst we shall not be dealing with the content of part 2, item 7, of these regulations it is worth reading as this deals with the requirements made of the supplier for protective multiple earthing. It follows therefore that this is the basis for the TN-C-S system for consumers.

Let's start by considering Part 6 of the Electricity Supply Regulations 1988, Supply to Consumer's Installations, as these will apply to installations with supplies derived from the Public Electricity Supplier (PES).

Regulation 25 refers to the requirements for the supplier's works on the consumer's premises and details the requirement, in item 2, for the standard of construction and installation of their works to be no less than that required of the consumer's installation by Regulation 27. It is Regulation 27 which tells us that the compliance of an installation with BS 7671 will give compliance with the requirements of the Electricity Supply Regulations 1988.

It is Regulation 25 which also calls for the supplier to provide a suitable fusible cut out or an automatic switching device as close to the origin as possible and, where the premises are not under the control of the supplier, this must be contained in a locked or sealed container. It is a further requirement that the supplier's conductors be identified to show the polarity of each conductor and, where appropriate, the phase rotation.

These requirements are typically dealt with as shown in Figures 1.1 and 1.2 with Public Electricity Supplier (PES) attached seals to their equipment to prevent unauthorised access.

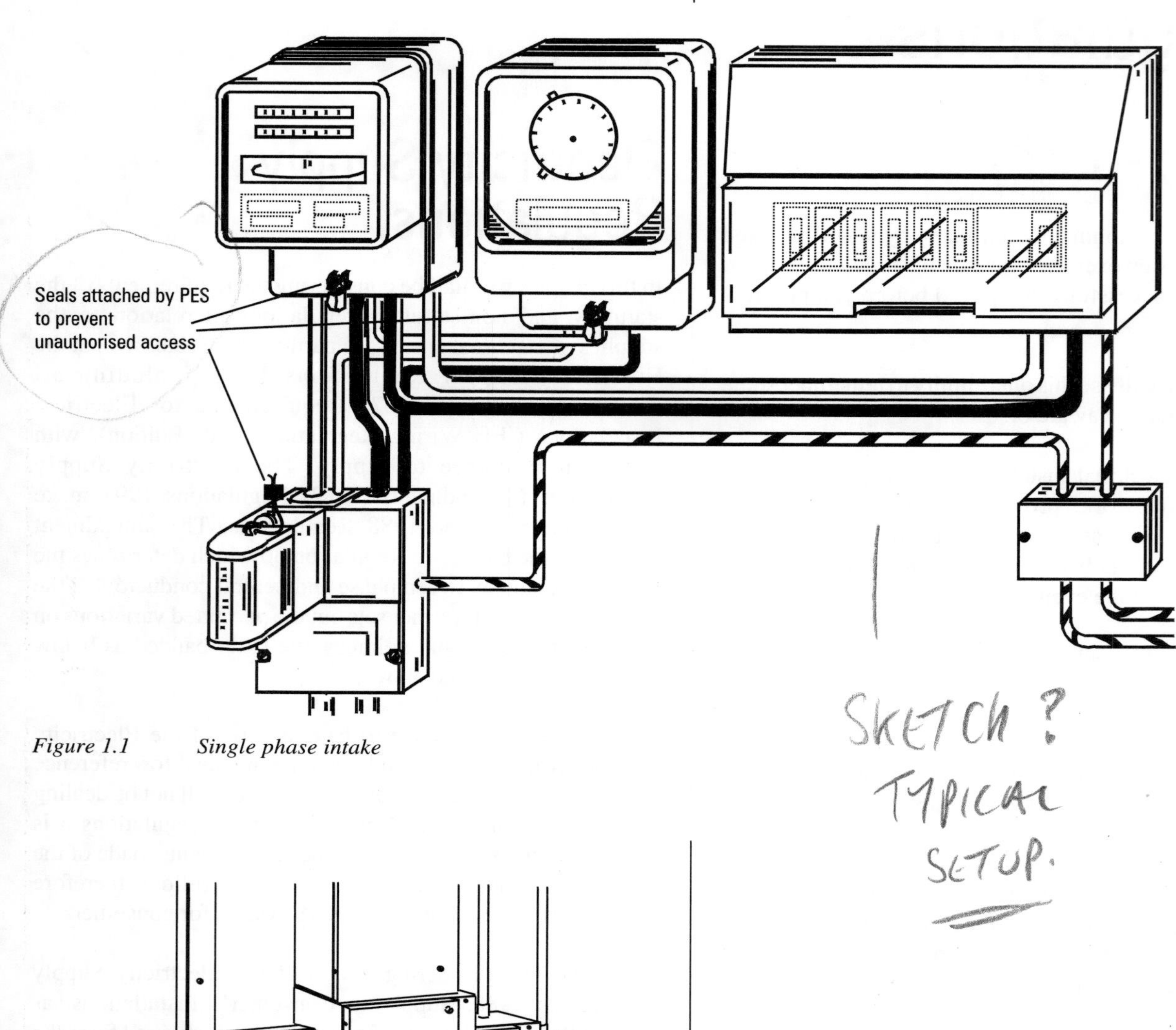

Figure 1.1 *Single phase intake*

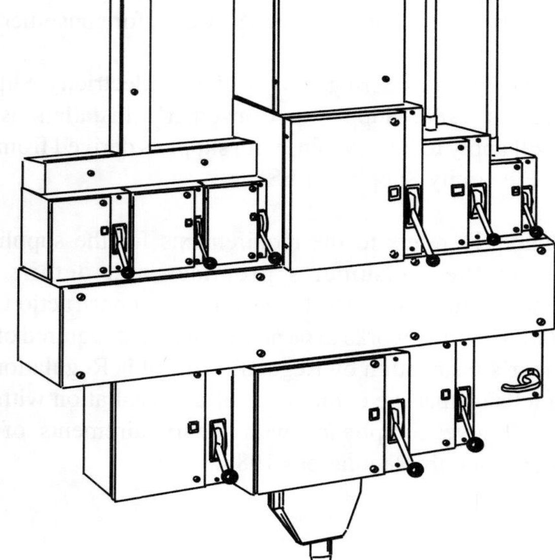

Figure 1.2 *Three-phase intake*

Remember

The fact that an installation complies with the requirements of the Electricity Supply Regulations 1988 does not necessarily give compliance with the requirements of BS 7671. However, an installation that complies with BS 7671 is deemed to give compliance with the requirements of the Electricity Supply Regulations 1988.

Try this

Locate one domestic and one industrial mains intake position and without infringing the requirements of either the Electricity Supply Regulations 1988 or the Electricity at Work Regulations 1989, and without interrupting the supply, sketch the layout of the supplier's equipment including cut outs and metering.

Make a note of the methods employed on the selected installation to meet the requirements for locked and sealed enclosures. Identify on your sketch the seals installed by the Public Electricity Supplier to secure their equipment at the intake position.

Regulation 26, although quite short, is of some importance particularly if our installation is to have more than one source of supply. This is the Regulation which gives rise to the PES requirements concerning the operation and connection of standby generators. This will, of course, include the switching, sequencing and control equipment. Schedule 3 of The Electricity Supply Regulations 1988 refers to the requirements and we can see that this includes the compatibility and synchronisation of supplies and the keeping of records of plant maintenance and failure.

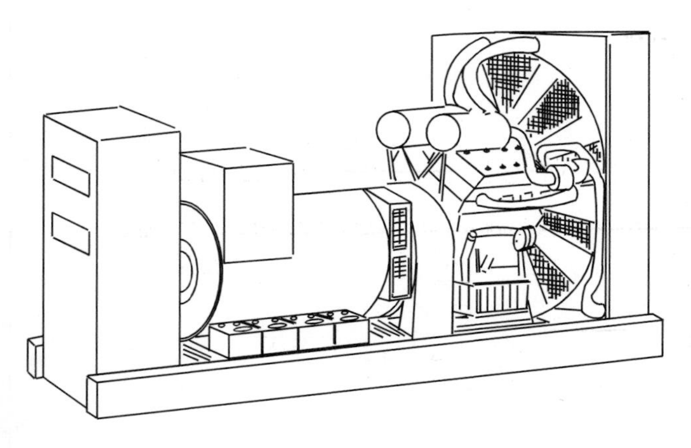

Figure 1.3 Diesel generator

Paragraph 3 of schedule 3 details the requirements for persons involved with the supply and part (a) makes reference to persons carrying out operations being "...authorised persons and competent to carry out such operations." From this we can deduce that consumers having their own generating plant, typically a diesel generator as shown in Figure 1.3, connected to the supplier's system, albeit through a switching arrangement, are obliged to comply with these requirements. Any staff employed to work on or operate this system must be competent to do so and a system of authorisation must be in place.

Regulations 27 to 29 inclusive are concerned with the supplier's duties and obligations in connection with supplies to consumers. They also detail the course of action that may be taken in the event of an installation failing to comply with the requirements of the Electricity Supply Regulations 1988.

Regulations 30 and 31 are concerned with the details of the supply that shall be made available to any person who can show reasonable cause for requesting this information.

Whilst we are not going to be considering Regulations 27 to 31 in this book, it is worthwhile reading through these and familiarising yourself with their requirements before continuing with this chapter.

Now we shall take a look at the requirements of Regulation 32 as this is relevant to the PES obligation for supply. We can see that once the supplier begins to supply a consumer, the conditions for maintaining that supply are quite onerous. This requirement gives rise to the supplier's network safeguard arrangements such as the provision of link boxes and ring mains. A simple example is shown in Figure 1.4. By careful linking and switching, supplies can be maintained in most instances where routine maintenance of the system is required. This system also provides the facility to rapidly bypass faulty cable and equipment thus minimising disruption in the event of an unscheduled incident. It is also customary for connections to the supply system to be made at LV level without isolation of the supply and as a result much LV mains jointing is done "live". However, this is carried out by suitable trained personnel using suitable tools, equipment and techniques to ensure compliance with The Electricity at Work Act 1989.

It is inevitable that there will be occasions when the interruption of the supply cannot be avoided, force majeure such as the hurricanes of 1987 or accidental cable damage being the most common causes. In such an event the supplier may be required to demonstrate to the Secretary of State, or the Secretary's authorised representative, that the requirements of the Electricity Supply Regulations 1987 have not been breached. This will be in the form of a declaration in accordance with Regulation 35 and this regulation sets down the circumstances in which this must be carried out. Details of the information to be provided are contained in schedule 5 of The Electricity Supply Regulations 1988.

There is one more regulation that we shall consider from the Electricity Supply Regulations 1988, before we move on, and that is Regulation 34 which deals with the notification of specified events. We shall not reproduce this regulation here but simply refer to the main requirements. Regulation 34 requires the supplier to give notification, to the Secretary of State, following the occurrence of an event as detailed in paragraph 2 of this regulation. The form of this notification is detailed in paragraph 3. Provision is made in paragraphs 4 and 5 for this to be done by the most expedient means in order to provide the earliest notification.

Schedule 4, of the Electricity Supply Regulations 1988, details the information to be supplied in the event of a specified event with pro forma schedules for parts I to IV inclusive.

The obligations made on the supplier by the Electricity Supply Regulations 1988 are quite demanding and place some responsibility on the supplier in connection with the consumer's installation. We are also bound by the Electricity Supply Regulations 1988 and as these are a statutory requirement we are legally obliged to ensure our installations comply. As we have already stated, one method of doing so is to ensure that we comply with the requirements of BS 7671, this will apply to all our installation work including the consumer's own supplies and distribution systems.

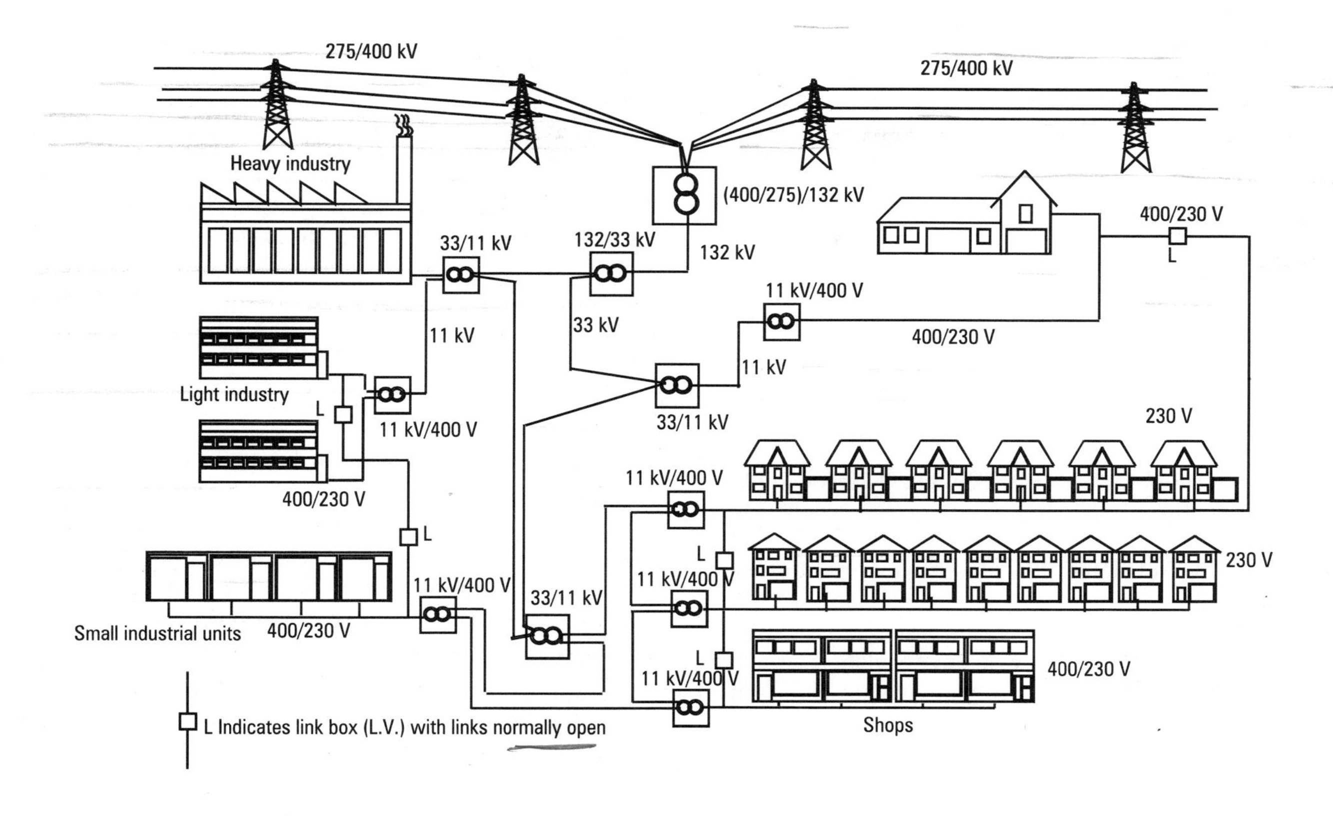

Figure 1.4 *Simple distribution system indicating the voltages used.*

We have considered a number of important parts of the Electricity Supply Regulations 1988 and it would be an advantage for you to take some time to familiarise yourself with their requirements before going any further with this chapter. Complete the questions in the Try this below before continuing.

Try this

1. List the specified events which may occur which will result in the Secretary of State being notified.

2. Is there a requirement for the supplier to keep and maintain maps of the underground supply cables within the Electricity Supply Regulations 1988 and, if this is required, under which regulation does it come?

We should be aware that the Public Electricity Suppliers go to considerable time, trouble and expense to ensure the competence of their authorised personnel. Training courses are provided and extensive procedures' manuals are in place for all those involved in work on their systems. Persons are only allowed to operate up to their level of authorisation and with the appropriate permits to work.

It is a requirement of us, as contractors, to ensure that we also are suitably competent and authorised for the work we are to carry out. If we are involved in operations on live substations and switchgear we must be suitably trained for this work and a permit to work system, under the control of the authorising engineer or operative, must be in place.

Our failure to ensure this is so will result in our contravening not only the Electricity Supply Regulations 1988 but also the Electricity at Work Regulations 1989 and the Health and Safety at Work Act 1974, all of which are statutory requirements.

We are also expected to comply with any regulations which may be issued by the Electricity Commissioners. This is in addition to, or in place of, those mentioned in the previous paragraph and we must always check to ensure whether there are any requirements, in addition to those we know, which may need to be taken into account. Most of the Public Electricity Suppliers have their particular requirements for TN-C-S (PME) systems and these may vary from area to area. We would be well advised to obtain a copy of the relevant PES booklet on their requirements before beginning a design or an installation in an unfamiliar PES area.

We shall be considering the relevant requirements of BS 7671 as we progress through this book and so we shall not be going into any depth in this chapter. As always it is our responsibility to make sure that we are aware of all the relevant standards and guidance material that are available for any aspect of our work. If we require additional guidance, we can contact such bodies as the British Standards Institute and the Health and Safety Executive.

Now attempt the following test yourself questions. Remember that the purpose of these questions is twofold, the first to check knowledge gained working through the chapter and the second to prompt you to further investigation into areas that are relevant to the topic we are considering.

Exercises

1. By the use of other documents, tables and reference material compile a list of the Electricity Suppliers who may be able to provide standards and guidance on the supply and distribution of electricity.

2. List those regulations within the Electricity Supply Regulations 1988 that will be of relevance to the electrical engineer in his design of a consumer's substation to enable energy to be purchased at 11 kV and then transformed and distributed by the consumer at 400/230 V.

3. Briefly describe the procedures that would be followed if an installation is found not to comply with the requirements of the Electricity Supply Regulations 1988 Regulation 27 and immediate action is required in the interest of safety.

2

Distribution

Revision questions:

1. Under what circumstances would the consumer be required to comply with the requirements of the Electricity Supply Regulations 1988, other than those requirements under Regulation 27 which relate to the electrical installation?

2. What information is the Public Electricity Supplier obliged to provide to anyone who can show reasonable cause for requesting the information?

3. When undertaking electrical installation work which is to form part of a TN-C-S system what additional information should be sought from the Public Electricity Supplier?

On completion of this chapter you should be able to:

- state the voltages used for transmission and distribution on the supply network
- determine likely fault levels
- state the advantages of ring and radial distribution systems
- list the information needed to select cables
- select the correct c.s.a. conductor, to supply given loads, from BS 7671
- calculate voltage drops for given circuits
- state the requirements for shock protection for circuits supplying both socket outlets and fixed equipment
- calculate the prospective earth fault current from given data
- check circuits for compliance with the requirements for shock protection
- state what is meant by thermal constraint and identify the criteria necessary to calculate compliance
- carry out calculations to find the minimum size of circuit protective conductor
- explain the need for load balancing
- apply diversity and determine maximum demand

In this chapter we shall consider the distribution of electricity by both the supplier and the consumer. Let's first have a brief recap on the supplier's transmission and distribution network. Transmission of electricity is the term generally applied to that part of the system that operates above 33 kV. Electricity is generated at up to 33 kV and is stepped up by transformers through several levels to a maximum of 400 kV to reduce the losses inherent in transmission over long distances.

Figure 2.1 serves as a reminder of a typical section of the supply companies' transmission and distribution network showing the appropriate voltages in each section.

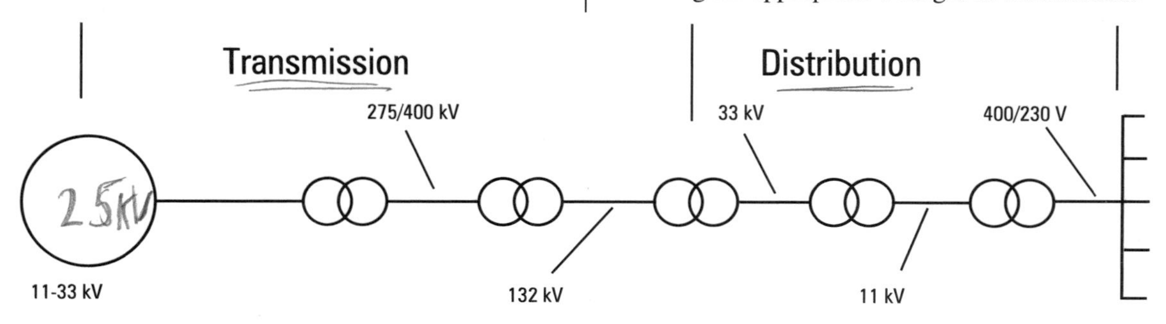

Figure 2.1

If the consumer receives energy at high voltage, the supply characteristics will have a similar bearing on the consumer's installation as those at low voltage. So we need to establish similar details in order to select appropriate switchgear and protection equipment at the origin of our installation. In the case of the high voltage supply the details of capacity will be given in kVA. The level of prospective fault currents will be required so that suitably rated main switchgear can be selected.

When considering high voltage distribution systems the fault levels are also going to be high and so it is common to consider "fault levels" when selecting switchgear and control equipment. The "Fault Level" is a measure of the electrical energy that will flow at the point of the fault. In most low voltage systems the fault current is the quantity considered, however, for high voltage systems we need to consider electrical energy. If we multiply the "fault current" by the nominal system voltage at the point of the fault, we arrive at the "Fault Level".

If we consider a fault current in the order of 13.1 kA on an 11 kV three phase distribution network then the Fault Level will be found by

Fault level	=	No of phases	×	Fault current per phase	×	Nominal system voltage to earth

So in the case above the fault level will be

$$\text{Fault level} = 3 \times 13100 \times \frac{1100}{\sqrt{3}}$$

$$= 250{,}000{,}000 \text{ VA or } 250 \text{ MVA}$$

This is considerably higher than the fault levels we encounter on the LV part of the system and the equipment installed must be suitable to cope with these levels.

It is worth mentioning, at this point, the effect of symmetric and asymmetric fault levels.

The value of 13.1 kA given in the above example is the symmetrical fault level and the Public Electricity Suppliers require their equipment to withstand this current for a period of 3 seconds.

Asymmetrical fault current is caused when a fault occurs on an inductive circuit, such as a transformer winding, when the voltage waveform is around zero.

$$\text{so } I(\text{peak}) = \sqrt{2} \times 13100 \times 1.8$$

$$= 33.3 \text{ kA}$$

This fault current will fall to symmetry level after approximately 5 cycles, around 0.1 of a second.

These fault levels are typical of the supply characteristics that the Public Electricity Supplier will provide to a consumer intending to buy electricity at high voltage. The Public Electricity Supplier will require all the equipment installed by the consumer to be suitable for the supply provided.

We shall consider some of the requirements for equipment and distribution at 33 kV and 11 kV in relation to the consumer later in this workbook. Larger consumers purchase electricity at these higher voltages and supply their own step down transformers and control gear.

Where a consumer operates over a large site, such as a car production plant or a large industrial complex, it is common for them to install their own distribution system and substations to supply particular parts of the site, often referred to as load centres. This considerably reduces the problems associated with voltage drop and power losses inherent in the distribution of electricity at 400/230 V. As the supply company does not have to provide and maintain the switchgear and transformer they offer an appropriately reduced tariff to large consumers, which reflects the supply company's savings on their system. Figure 2.2 shows the entrance to a large industrial consumer's premises.

Figure 2.2 Large industrial consumer's premises

We saw in the previous chapter that one of the most important considerations is the reliability of the supply. If consumers purchase energy at, say 11 kV, and then install their own transformers and so on they must safeguard against the loss of supply on their part of the system. If we consider the simple line diagram in Figure 2.3 we can see that a break at any point after the intake position leaves all the equipment downstream of the break in supply.

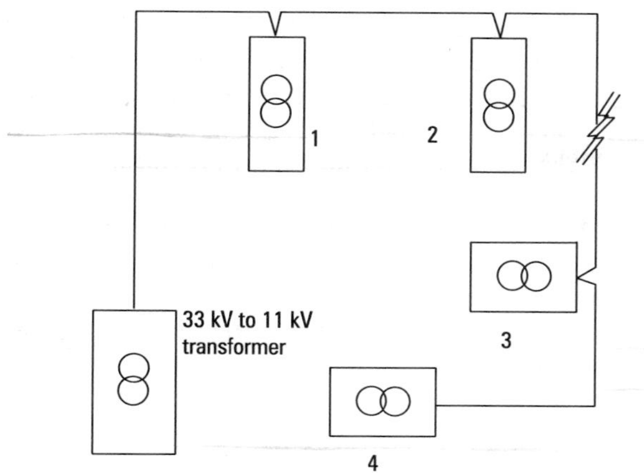

Figure 2.3 *The break between units 2 and 3 renders units 3 and 4 without supply*

By installing the 11 kV supply as a "ring main", as shown in Figure 2.4, we can see that units 3 and 4 could be supplied from the source if we could isolate the damage between 2 and 3. The use of a switch arrangement as shown in Figure 2.5 enables us to do just that.

Note: The term "Ring Main" is used here in the correct context. A similar circuit arrangement for 13 A socket outlets should always be described as a "ring final circuit", often shortened to ring circuit, but never as a ring main.

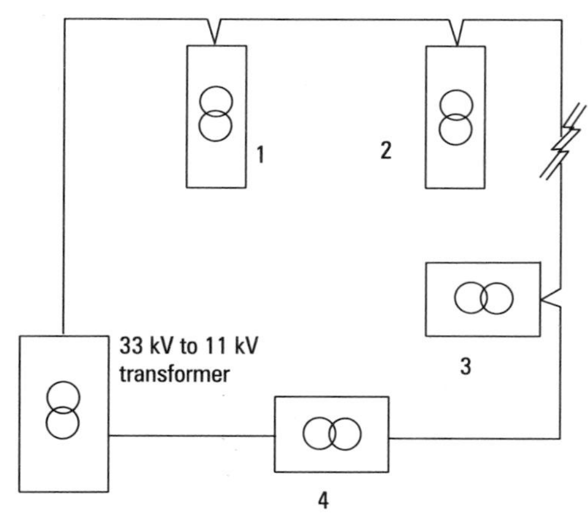

Figure 2.4

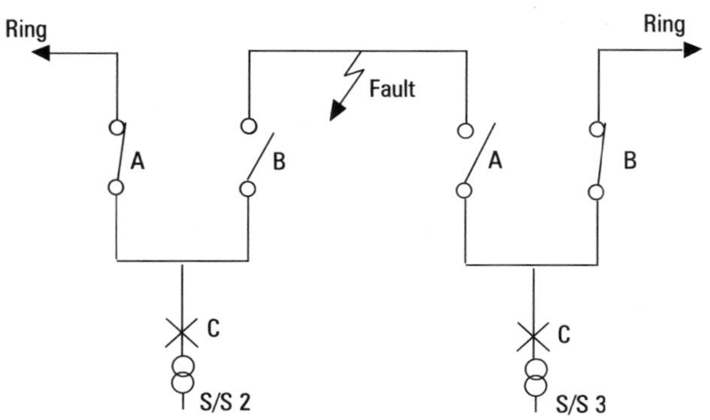

Figure 2.5 *Switch arrangement:*
Open switch B at S/S 2
Open switch A at S/S 3
This will isolate the faulty cable section.

This is a good time to consider radial and ring main distribution systems and look at the advantages of each in turn beginning with radial distribution.

Radial distribution

The line drawing shown in Figure 2.6 is for a radial distribution network that has a low initial cost with a single primary substation and a radial feeder looped to the four secondary substations. The radial is controlled by a single circuit breaker and the secondary transformers by switches which are capable of making under fault conditions and breaking under load conditions. The transformer output is protected by a single circuit breaker.

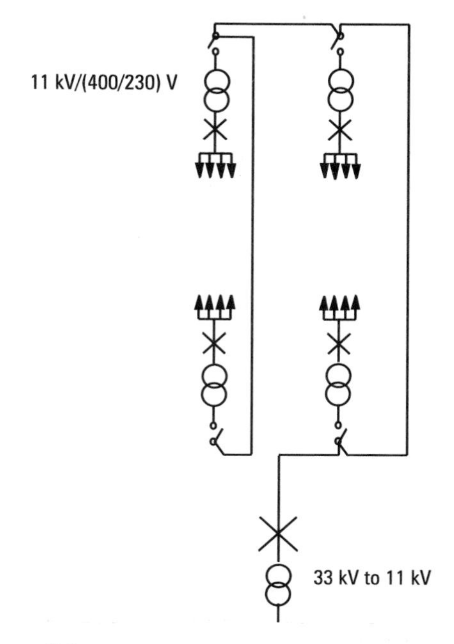

Figure 2.6

With this system any transformer or a section of the low voltage system may be isolated without interruption of the supply to the other substations. However any work on the secondary substation transformer or switchgear means that the whole of the LV must be isolated for the duration of the work. Any work required on the primary substation transformer or circuit breaker will require the interruption of the supply to all the secondary substations. Any problem with the radial supply will cause disruption to all substations downstream.

> *Remember*
>
> 
>
> An electrical system is often likened to a river with current flowing from the source downstream to the final load. If we consider a point on the system then any part of the system which is on the supply side of that point is said to be UPSTREAM, any part of the system that is to the load side of that point is then said to be DOWNSTREAM.

Ring distribution

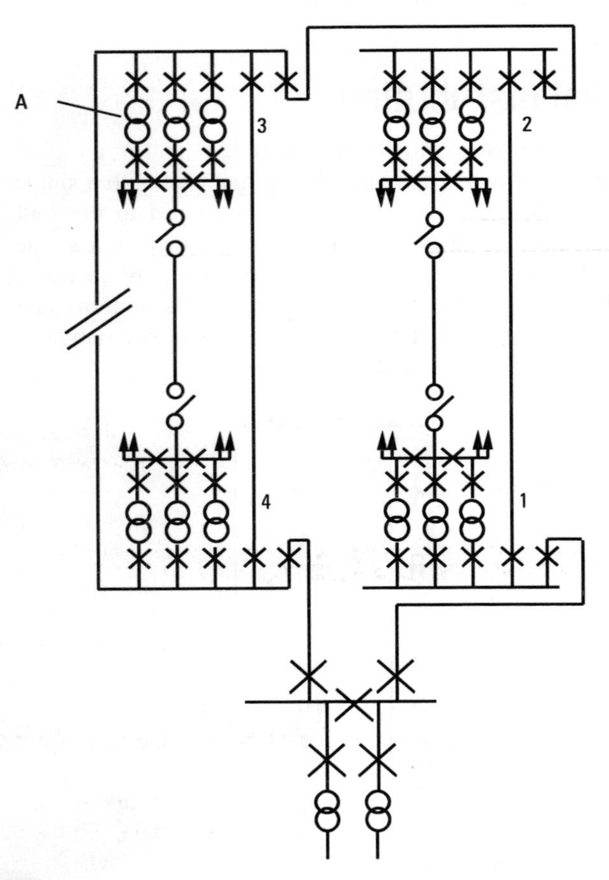

Figure 2.7

Figure 2.7 shows an industrial ring distribution system. This has been arranged to provide continuation of supply to the entire system during maintenance. In the event of a fault occurring on the system the supply can be restored quickly, the exception being the total failure of the supply to the primary sub station. This system has a considerably higher initial cost but provides a secure supply under almost all conditions and allows the system to be maintained without interrupting the supply.

In the event of an item of equipment becoming faulty the engineer can arrange for the system to be switched to allow the supply to be restored prior to repair. The faulty item can then be repaired or replaced without further disruption.

Within the diagram the open isolators and linking cables do not form part of the system. The function of these components is to allow the engineer to maintain the supply during maintenance or repair. By the opening and closing of the relevant isolators these tasks may be carried out without the need to interrupt the supply to any part of the system.

It is not uncommon to supply alternative cables both on the upstream and downstream side of transformers, where possible, to allow the engineer to maintain the supply to all parts of the system during maintenance. With very large consumers it is a common practice to have two or three identical transformers connected in parallel all of which are capable of supplying the total load required. These transformers may then be run at only half or one third load but provide the facility for any one to be shut down in the event of a fault or for maintenance. Alternatively the transformers may be regularly cycled to share time on load.

The use of ring mains and selective switching arrangements, coupled with the provision of parallel transformers, enables the engineer to maintain the supply to all areas during routine maintenance. In the event of a breakdown or fault the disruption of supply will be kept to a minimum, resulting in minimum loss of production. The more sophisticated the system the more reliable the supply. The client will need to balance the initial and maintenance cost of such a system against the probable cost of supply failure.

Try this

The diagram below shows the details of sub stations 3 and 4 from Figure 2.7. It is necessary to isolate transformers A and B in sub station 3. The system is designed to operate with at least two transformers in each sub station in operation to supply the connected load. At this time it is transformers A & B, in each substation, which are operational. Identify the isolation devices which need to be opened or closed and the sequence in which they need to be operated in order to isolate the two transformers without interruption of supply to the installation or overloading the system.

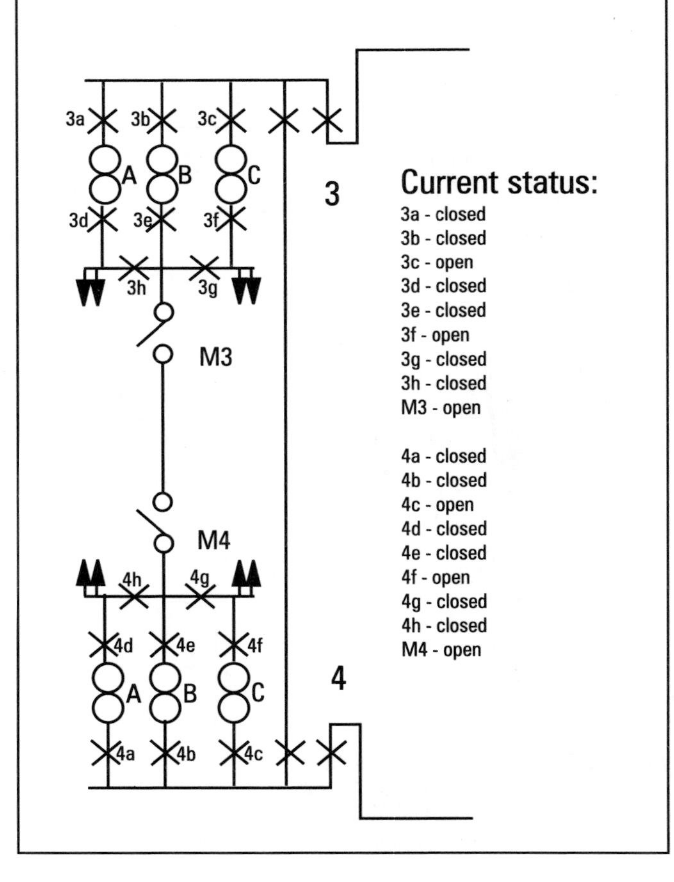

So far we have discussed the distribution system for a consumer in similar terms to that employed by the supply company. In many cases the ring main cables will not be placed underground and indeed may not be installed outside of a single building. Such installations often employ busbar trunking as the ring main with suitable switchgear located within substations sited around the building.

The installation of a busbar or cabled ring or radial mains within a consumer's premises is subject to the same criteria as we have considered and the consumer must determine the extent to which the cost of installation compares to the possible loss of production. Remember that the system must be maintained on a routine basis and this will require the isolation of equipment.

By using a suitably sized distribution system, with appropriate linking facilities, a compromise between cost and supply reliability can generally be reached. Should the maintenance of supply be essential, as in the case of computer equipment, then the consumer will usually elect to install an Uninterruptible Power Supply (UPS) and possibly a standby generator in addition to the arrangements mentioned so far.

Generally the LV part of the system will also have a distribution system. For consumers purchasing electricity at high voltage this will usually emanate from the secondary substation. For consumers with low voltage intakes the LV distribution usually emanates from the intake position.

Cable sizing

This is a good time to consider the requirements for the selection of appropriately sized distribution cables. The process of sizing LV cables to meet the requirements of BS 7671 has been covered elsewhere in this series. However, it is worth reviewing the sequence and stages involved in the selection process.

In order to size the conductor and protective device correctly we need to consider the requirements in an appropriate sequence. The following list details the considerations and calculations which have to be made and the sequence in which they should be approached.

1. Determine the connected load for the section of the installation to be supplied.
2. Apply diversity as appropriate to determine the maximum demand requirement.
3. Determine the full load current and design current for the maximum demand.
4. Determine the rating of the overcurrent protection device.
5. Determine the factors that apply.
6. Determine the minimum current carrying capacity of the conductor.
7. Determine the reference method.
8. Select the cable size from the appropriate table based on current carrying capacity.
9. Ensure cable selected complies with the voltage drop constraint.
10. Determine the maximum value of earth fault loop impedance from the tables.
11. Calculate the actual value of Z_s for the circuit.
12. Ensure the circuit complies with the shock protection constraint.
13. Calculate the earth fault current.
14. Establish the disconnection time from the time/current characteristics.
15. Calculate the minimum cross sectional area of the circuit protective conductor.
16. Verify that the circuit complies with the thermal constraints.

Having considered the sequence for determining the requirements we shall review some of these aspects in more detail beginning with

Maximum demand

The first point we need to establish is the maximum demand that will be required for the installation.

For small installations this may be done by the use of actual connected load values at an early stage in the design.

Larger installations and commercial premises are often assessed on a rule of thumb basis in the initial stages. This is because the actual loads of machines, for example, will not be known until later. Designers use a method of approximation based on the use of the building and floor areas involved. A storage warehouse, for example, would normally require a relatively low level of lighting and a minimal power load for a fairly large floor area. A small engineering unit may, by contrast, have a high level of lighting and a considerable power requirement for a much smaller floor area.

By using a table giving the load requirements per square metre for various types of building uses and the floor area to be served the designer can quickly approximate the electricity load requirements for a proposed building.

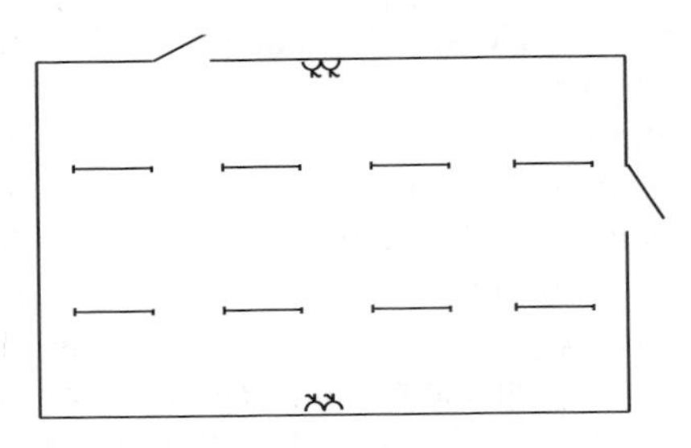

Figure 2.8 *Plan of store room*

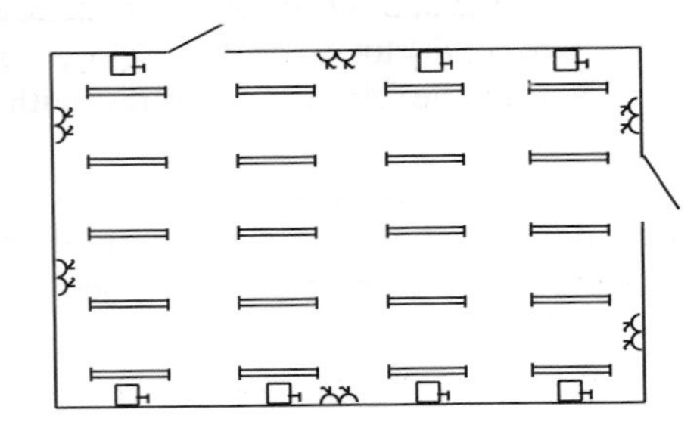

Figure 2.9 *Plan of workshop*

The above rooms are of the same area but the electrical load on the store room is far less than that of the workshop.

Determining maximum demand

We will consider the maximum demand of an installation to be the power required if every piece of equipment installed was switched on.

Let us consider a simple example in a domestic installation with the following equipment installed:

	Total
1 × cooker rated at 45 A	45 A
2 × ring circuits rated at 30 A	60 A
1 × immersion heater rated at 15 A	15 A
2 × lighting circuits rated at 5 A	10 A
	130 A

We have established a total load of 130 A for our domestic installation but it is unlikely that the supply company will install a service larger than 100 A for this installation. This is the result of applying a diversity to our total connected load. This is a method by which an estimate of the total required load is established based on the assumption that all the equipment connected will not all be used at the same time. We shall not be considering this in detail at this time as we are concerned with the selection of cables to supply specific load currents.

Now that we know the information that we need to begin our selection process we can do a quick refresher in calculating the current requirements of given loads. Remember that for a resistive load our power formula is

$$\text{power} = \text{voltage} \times \text{current}$$

If our load is inductive then we must include the power factor in this calculation and so our formula becomes

$$\text{power} = \text{voltage} \times \text{current} \times \text{power factor}$$

If the load connected is given in watts or VoltAmperes then by rearranging the formula we can find the current that will be drawn. For our resistive load the current drawn will be given by

$$\text{current} = \frac{\text{power}}{\text{voltage} \times \text{power factor}}$$

If we now consider a balanced three phase load then the power required for a resistive load will be given by the formula

$$\text{power} = \sqrt{3} \times U_L \times I_L$$

and for an inductive load

$$\text{power} = \sqrt{3} \times U_L \times I_L \times \text{power factor}$$

By rearranging these formulae as before we can determine the line current required for a balanced three phase load.

$$\text{line current} = \frac{\text{power}}{\sqrt{3} \times U_L \times \text{power factor}}$$

> *Remember*
>
> 
>
> Power factor is, in fact, the cosine of the phase angle between the voltage and current, known as cos θ, so the formula may appear as
>
> $P = UI \cos\theta$ for single phase and so on.

Let's consider a simple example.

A circuit is to supply an electric heater at 230 V 50 Hz and this heater is rated at 3.5 kW. What will be the current drawn from the supply if the power factor is unity (1)?

> *Remember*
>
> 
>
> Purely resistive loads will have a power factor of unity.
>
> We usually consider a heater of this size to be a resistive load with a power factor of 1.

Power (watts) = voltage (volts) × current (amps) × power factor

$$\text{current} = \frac{3500}{230 \times 1}$$

$$= 15.217 \text{ amperes}$$

This is the "Design Current" for this heater and is given the symbol I_b. It is the current that will be drawn by this heater under normal operating conditions.

Remember

I_b is the design current of a circuit or load and is the current drawn from the supply under normal conditions.

If we now apply the same principle to a balanced three phase load of, let's say, 15 kW at 400 V, 50 Hz and a power factor of unity (1) then we have:

$$\text{Power} = \sqrt{3} \times U_L \times I_L \times \text{power factor}$$
$$15000 = \sqrt{3} \times 400 \times I_L \times 1$$

so the line current will be

$$I_L = \frac{15000}{(\sqrt{3} \times 400 \times 1)}$$

Calculate the terms inside the brackets before you divide.

$$= 21.65 \text{ amperes}$$

So for this balanced three-phase load the design current I_b will be 21.65 amperes per phase.

Try this

1. A single phase load of 2.5 kW is to be supplied at 230 V 50 Hz. If the load has a power factor of 0.9 what will be the design current of the circuit?
   (Remember to calculate the terms inside the brackets before you divide.)

2. We are to supply a balanced three-phase load at 400 V 50 Hz. If the power required is 25 kW at power factor of 0.8 what is the design current?

Try this

Complete the following:

The __________ __________ of an installation will be the power required if every piece of equipment installed was switched on.

If __________ is applied to the total required load this is based on the assumption that all the equipment will not be used at the same time.

The power formula for a resistive load is:

power =

The power formula for an inductive load is:

power =

For a resistive load we can calculate the current drawn by using the formula:

current =

For a balanced three-phase load the power required will be given by using the formula:

power =

Purely resistive loads will have a power factor of __________.

The current drawn from the supply under normal conditions is the __________ current and the symbol used for it is __________.

Having calculated design current for single and three-phase circuits we shall consider the selection of the correct size of cable to supply the various parts of the system. This will include some of the cables that go to make up the consumer's installation. For larger installations where the consumer purchases energy at 11 kV or above it will be necessary to determine the size of the consumer's own supply cables.

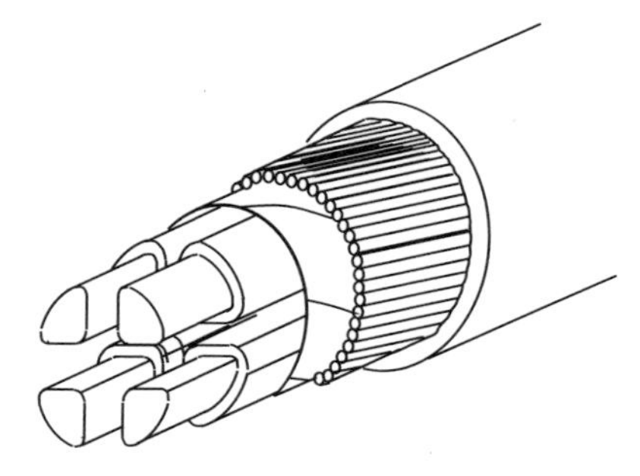

Figure 2.10 *Typical distribution supply cable*

Conductor selection

In some cases we may find that the installation imposes conditions that cause us to reconsider our cable sizing. We may then need to choose an alternative method, or system, in the interests of practicality, economy or ease of installation. The first part of the process is to determine the design current of the circuit, remember that this is the current that the load will require under normal conditions.

We shall begin by looking at a fairly simple circuit supplying an electric heater. For this exercise we will assume that the heater has a total power requirement of 3 kW when supplied at 230 V 50 Hz.

Design current

Using the data we are given (remember that being a resistive load we will consider that the power factor is unity) we get a required current of:

$$\text{Required current} = \frac{3000}{230 \times 1}$$

$$= 13.04 \text{ amps}$$

So if 13.04 amperes is the current required for the heater to operate normally then, logically, the fuse or circuit breaker used to protect the circuit must be capable of carrying this current, without damage or deterioration, for an indefinite time.

Now protective devices are manufactured in a standard range of sizes, so the next task is to determine the appropriate size of device to be used. The actual rating will depend on the type of device used so we must decide on the type before we can select the rating. Some of the types and ratings are shown in Table 2.1.

Table 2.1

Protection type	Current rating
BSEN 60269-1:1994 (BS 88)	2 4 6 10 16 20 25 32 40 50 63 80 100 125 160 200 250 315 355 400
BSEN60269-1:1994 (BS 88)	355 400 450 500 560 630 710 800
BSEN60269-1:1994 (BS 88)	2 4 6 10 16 20 25 32
BS 1361	15 20 30 40 45 50 60 80
MCB BSEN 60898/BS3871	Types B & C (2 and 3) 6 10 16 20 32 40 50 63

Remember

The protective device that we install is to protect the **cable** supplying the equipment, **not** the equipment itself.

In this case we are going to use a BS 1361 type fuse and by reference to the manufacturer's data or in Table 2.1 we can see that the nearest size of fuse is a 15 A. This is the nominal rating of the fuse, that is to say the current that the fuse will carry for an unlimited period of time without deterioration. It is referred to as I_n and it is important that we remember that the rating of the device I_n is equal to or greater than the design current for the circuit I_b.

This may be written as $I_n \geq I_b$

This rule must always be applied so for example if the design current of a circuit was 20.5 A we could not use a 20 A BS1361 fuse for protection and we would have to go to a 30 A device.

> *Remember*
> Always round up when selecting the rating of the protective device **NEVER** round down.

So for our heater we know that we need to install a 15A fuse. The next stage is to determine the current carrying capacity of the cable and so select a suitable size of conductor. The first thing to establish is the minimum current carrying capacity of the cable required and we shall refer to this as I_t.

However, before we can do this we must give some consideration to the conditions under which the cable will be operating and the method of installation.

Whatever type of material we use for a conductor it will have some resistance. We know that the factors, which affect the value of this resistance, are

- the material from which the conductor is made
- the cross sectional area (c.s.a.) of the conductor
- the length of the conductor

If we pass current through a resistor, we produce heat and a voltage drop occurs across the installation.

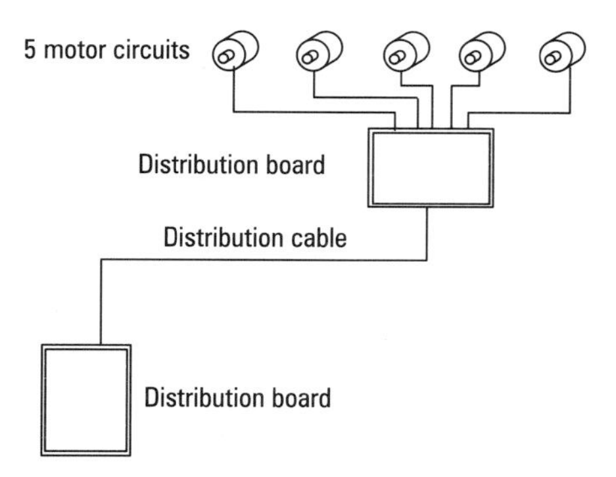

Figure 2.11 *When all of the motor circuits are loaded the voltage drop on the submain cable should not be excessive*

A rise in temperature will also produce an increase in the resistance of almost all metallic conductors. We must try to keep the heat produced in the conductor to a minimum during operation. To enable us to do this we must consider the types of location and conditions which will affect the heat that a cable can dissipate (give off).

The main points are:

- ambient temperature
- grouping
- thermal insulation
- type of protection device used

Ambient temperature

This is the temperature of the surroundings of the cable. It is often the temperature of the room or building in which the cable is installed. Now, the hotter the surroundings the less heat the cable will be able to give off. We put food in a warm oven to keep it hot. If the surrounding temperature is low then the heat dissipated is greater and so the cable would "run cooler", and so could give off more heat.

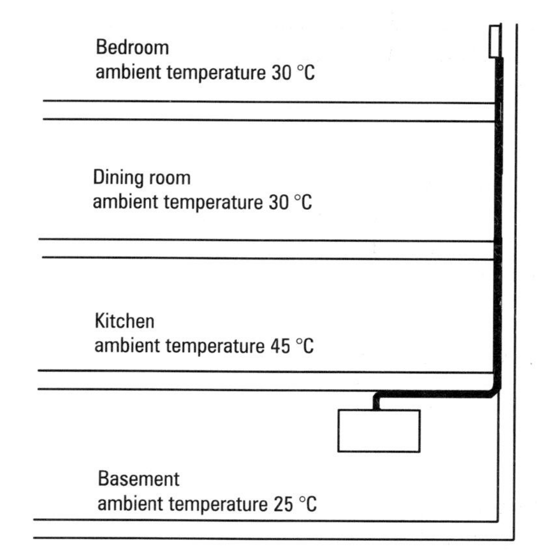

Figure 2.12

Grouping

If a number of cables are run together, they will all produce heat when they are carrying current. The effect of this is that they reduce the heat dissipation by each other. The same effect is used by groups of animals huddling together to keep warm. If we keep the cables separated then this effect will be minimised.

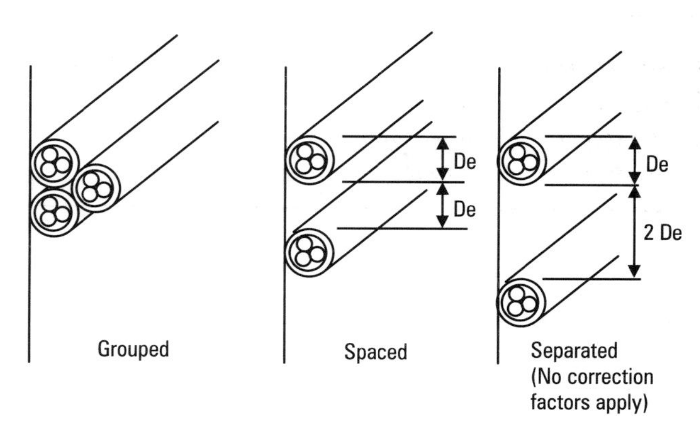

Figure 2.13 *De = the overall cable diameter*

Thermal insulation

This has a similar effect to wrapping a cable in a fur coat on a hot summer's day and the heat produced within the cable cannot escape. In terms of electrical installations there are two conditions to be considered, these being

- cables in contact with thermal insulation on one side only, such as in Method 4 in the scheduled methods of installation of cables in BS 7671
- cables completely surrounded by thermal insulation

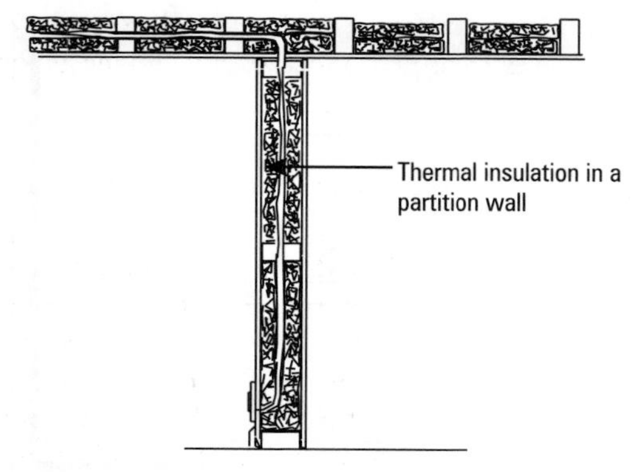

Figure 2.14

> *Remember*
>
> 
>
> When we use a device to protect a cable, it usually operates when too much current is drawn through it. Excess current will produce more heat within the cable and, unless the device can disconnect quickly enough, damage may be caused to the cable insulation. Under extreme conditions the insulation may catch fire.

Applying factors

Now that we have seen the situations and conditions that can cause too much heat in a cable what can we do to prevent this occurring?

Whilst we cannot prevent some of these conditions, we can take some precautionary steps. Modification of the route to avoid a particular area, or installing cables with adequate space between them, are two examples.

This will not always be possible and so we must have some other approach to ensure a safe installation. To do this we use a system of FACTORS, one for each condition. The purpose of these is to cause us to use a larger c.s.a. of conductor to reduce the resistance and so reduce the heat generated within the cable.

Let's look at the factors we use and see how they are applied. These factors are given in BS7671 so we shall also consider where they are to be found within that document.

Ambient temperature

The tables used for cable selection will be based on a particular ambient temperature. In the case of BS7671 this is 30 °C so any cables installed in an ambient temperature above this will need to be adjusted as they cannot give off as much heat. A set of factors for these conditions are given in BS7671in Tables 4C1 and 4C2. You will notice that the type of insulation used also has an effect on the factors given in the tables. This factor is referred to as C_a.

> *Try this*
>
> 
>
> Complete the table below using a current copy of BS7671.
>
> Correction factors for ambient temperature

Type of insulation	Ambient temperature °C							
	30	35	40	45	50	55	60	65
60 °C rubber (flexible cables only)								
General purpose PVC								

Table 2.3

> *Try this*
>
> 
>
> Complete the table below using a current copy of BS7671.
>
> Factors for ambient temperature where the overload protective device is a semi-enclosed fuse to BS 3036.

Type of insulation	Ambient temperature °C							
	30	35	40	45	50	55	60	65
60 °C rubber (flexible cables only)								
General purpose PVC								

Table 2.4

Grouping

The factors used for grouping are contained in Table 4B1. By reference to 4B1 you will see that the method of installation also has some bearing on the factor to be used. It is important to remember that these factors are applied to the number of circuits or multicore cables that are grouped and not the number of conductors, an important point when we are installing in conduit or trunking. This factor is known as C_g.

Try this

Complete the table below using a current copy of BS7671.

Correction factors for grouped cables

Reference method of installation		Correction factor (C_g)							
		Number of circuits or multicore cables							
		2	3	4	5	6	7	8	9
Enclosed (Method 3 or 4)									
Single layer multicore on cable tray (11)	T*								
	S*								

Table 2.5

*T – touching

*S – spaced, "spaced" means a clearance between adjacent surfaces of at least one cable diameter (De).

Remember

Grouping is the number of circuits or multicore cables **NOT** the number of conductors involved.

Thermal insulation

This is dealt with in two ways. First if the cable is surrounded by thermal insulation then the factor that must be applied is a constant value of 0.5 (or for cables up to 10mm ^{2} over short distances as shown in Table 52A), this factor is known as C_i. If the cable to be installed is in contact on one side only with the thermal insulation then this situation is dealt with by using the tables for this installation method. We shall consider this later as we do not have to apply a factor for this particular situation.

Type of protective device used

As we noted earlier, this factor is really dependent on the speed and reliability of the protective device. It is sufficient to say that the only device that does NOT disconnect in sufficient time is the BS3036 semi-enclosed rewireable fuse.

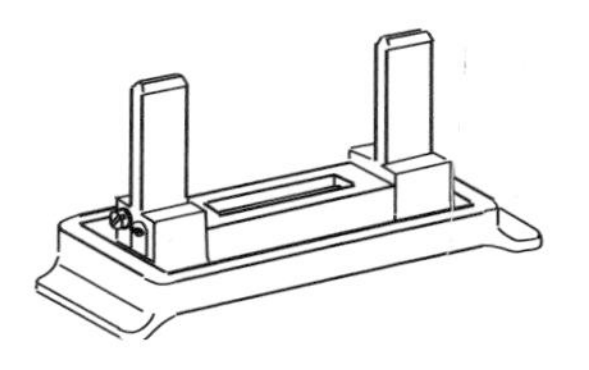

Figure 2.15 BS 3036 semi-enclosed rewireable fuse

If we select this particular device then we must **ALWAYS** use a factor of 0.725 when calculating current carrying capacity. This value is derived from Regulation 433-02-03 which states that for this type of device I_n should not exceed 0.725 × the lowest cable current carrying capacity, giving us our factor C_f for the fuse.

$$I_n \leq (0.725 \times \text{lowest cable current carrying capacity})$$

So how do we apply these factors to determine the size of conductor that we require? You will notice that nearly all of these factors are less than 1. We also know that the current carrying capacity of the cable must be equal to or greater than the rating of the fuse that protects it. The purpose of applying the factors is to ensure that the conductor is large enough to carry the current without excessive heat being generated and if the dissipation is affected by adverse conditions then the only way in which this can be resolved is by increasing the size of the cable.

When a protective device operates it generally relies on an overcurrent to do so. If a device is rated at 20 A then this is the current that it will carry for an indefinite period without deterioration. It follows then that the device will only register current beyond its rated value as overcurrent so we must use the rating of the device to carry out our calculations to find the value of I_t.

Remember

It is the CURRENT RATING of the protective device, I_n, that is used in calculations to determine I_t.

If more than one of our conditions for the application of factors exists then we must consider the worst case. If, as an example, a cable runs through an area of high ambient temperature then it is grouped with several other cables at another location and is finally run totally enclosed in thermal insulation at another point then we consider all the factors and apply the most onerous. However, if more than one condition applies at a single location then we must apply all those factors.

To establish the minimum value of I_t for any circuit we must divide the current rating of the protective device by all the factors. If any factor does not apply, we give it the value 1.

So $$I_t \geq \frac{I_n}{(C_a \times C_i \times C_g \times C_f)}$$

Where C_f is the factor for a BS3036 fuse.

If no factors need to be applied then

$$I_t \geq \frac{I_n}{(1 \times 1 \times 1 \times 1)}$$

and in this case

$$I_t = I_n$$

Remember

I_t is the tabulated value. This will not necessarily be the same as a calculated result.

I_t is always equal to or greater than I_n or

$$I_t \geq \frac{I_n}{\text{factors}}$$

Example

If the cable supplying our heater is run through a roof space with an ambient temperature of 35 °C, and if this is the only factor that applies, then we can calculate the value of I_t by simply dividing the fuse rating by the factor for an ambient temperature of 35 °C.

So we have

$$I_t \geq \frac{15\text{ A}}{C_a \times 1 \times 1 \times 1}$$

So we turn to Table 4C1 (remember we are using a BS1361 fuse). The first thing that we find is that we need to know more details about the circuit that we are to install. In this case we need to know the type of insulation that we are to use. In this example we shall use a PVC PVC cable as it is a normal domestic installation. This means we have the value of 0.94 for general purpose PVC.

So our value for I_t will be

$$I_t \geq \frac{15\text{ A}}{0.94} \qquad \geq 15.957\text{ A}$$

You can see that the current carrying capacity has increased to almost 16 A as a result of the effect of a higher ambient temperature.

I_t for this situation is 15.957 A

If the same cable is to be run through the roof space but is grouped with 3 other cables, in an ambient temperature of 35 °C, and it is protected by a BS3036 type fuse then we must apply all these factors to establish I_t so we get:

$$I_t \geq \frac{15\text{ A}}{C_a \times C_g \times C_f \times C_i}$$

(C_f for the BS3036 fuse is 0.725)

$C_a = 0.97$ from Table 4C2

When we come to establish the value of C_g, using Table 4B1 BS 7671, we find that we need yet more information about the circuit that we are to install. This time we need to know the method of installation. This is the method of wiring, so we need to know how the cable is to be installed. The method used in this example is to run the cables clipped direct to the surface of the building so that is the method that we shall use.

$C_g = 0.65$ from Table 4B1

We must remember to use the total number of cables that are bunched together. For this example this is the 3 other cables plus the 1 cable that we are to install giving a total of 4.

C_i does not apply in this case as the cable is not in contact with thermal insulation at any point on the route, so $C_i = 1$.

So to complete the calculation we have

$$I_t \geq \frac{15}{0.97 \times 0.65 \times 0.725 \times 1}$$

$$I_t \geq 32.81\text{ A}$$

As you can see, this has a considerable effect on the current carrying capacity of the conductor required. In practice it would be a more sensible idea to run the additional cable through the roof separate from the other cables and consider an alternative protective device. This would then reduce the number of factors that need to be applied.

Try this

Using the cable from the above example we are going to install it for the heater so that it is clipped separately from all the other cables.

Calculate the minimum value of I_t for this situation if all other factors remain unchanged.

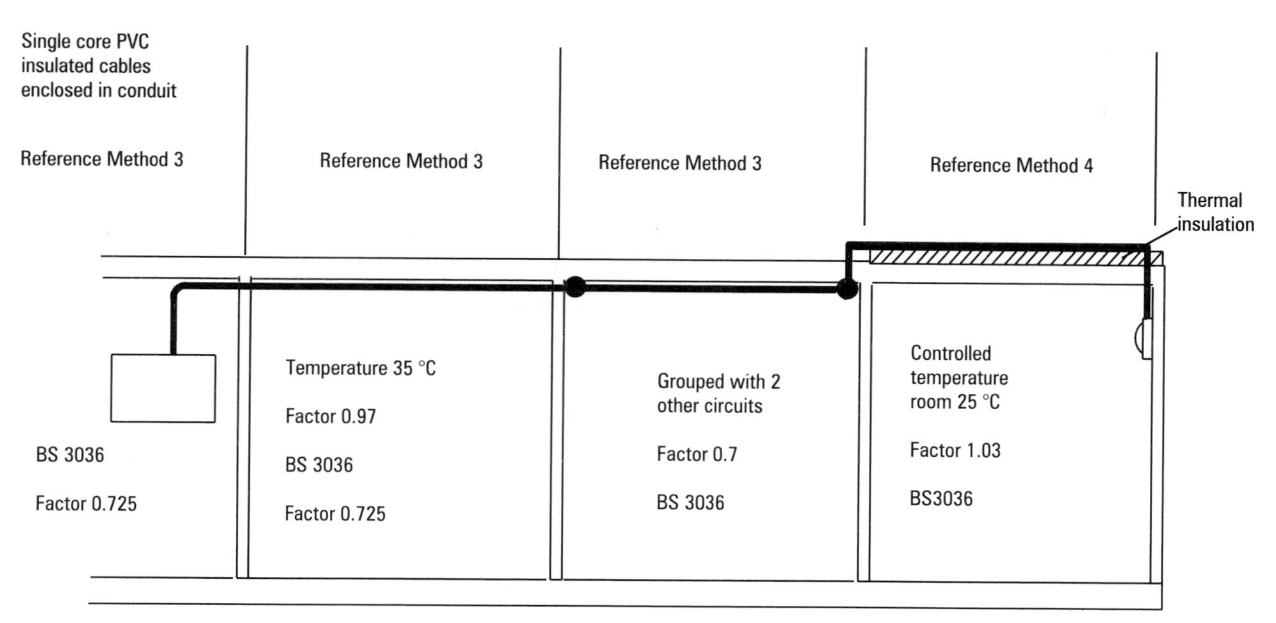

Figure 2.16 *The most onerous conditions in the above diagram are where the cables are grouped with two other circuits. This gives a factor of 0.725 × 0.7 = 0.5075. (The factor for the BS 3036 fuse applies throughout the whole length of cable run.)*

There are going to be situations where a number of factors will need to be applied but not all at the same point along the route. To cater for these situations we take the worst condition that exists along the route and size our cable to those requirements. If it complies with the worst conditions then it must be satisfactory for the rest.

Before we can carry out this exercise we must have the following information:

- the type of protective device that is to be used
- the type of cable that is to be installed
- the method of installation to be employed
- the ambient temperature that is likely to be present along the route
- any grouping that may occur along the route
- any thermal insulation that may be encountered on the way

This will provide us with the most onerous conditions that we will encounter. We can, of course, reconsider our route, method of installation, type of cable or protective device in order to produce the most economic and practical installation we can.

Remember

We may need to re-route and so on to reduce the numbers of factors that have to be applied, thus preventing the cable size becoming impractical.

Try this

The following symbols are used in BS7671. Using Appendix 4 or Part 2 in BS 7671 describe what these symbols mean:

I_t

I_z

I_b

I_n

C_a

C_g

C_i

C_t

Try this

1. State:
   (a) the effect on the current carrying capacity of a cable that is totally enclosed in thermal insulation

   (b) what effect this will have on the size of the cable required and why.

2. State the effect that grouping cables together has on their current carrying capacity and why this is so.

3. A PVC/PVC cable is to supply a load of 2 kW at 230 V, 50 Hz and unity power factor. If the circuit is to be protected by a BSEN 60269-1:1994 type fuse and the cable will be grouped with 2 other cables throughout the run what will be the minimum value of I_t for this circuit?

Having calculated the minimum current carrying requirements for conductors to supply a given load we will now consider the selection of suitably sized conductors, to supply loads under defined conditions, and ensure compliance with the voltage drop constraints.

The first stage is to determine the minimum current carrying capacity of the conductor.

To do this we will use a simple circuit which has no factors to be applied.

A circuit is to supply a load of 16 A at a distance of 20 m from the supply intake position. The supply is 230 V 50 Hz and the circuit is to be protected by a BS1361 type fuse. The circuit is to be installed using PVC/PVC cable clipped direct to the surface and the conditions are such that no factors need to be applied.

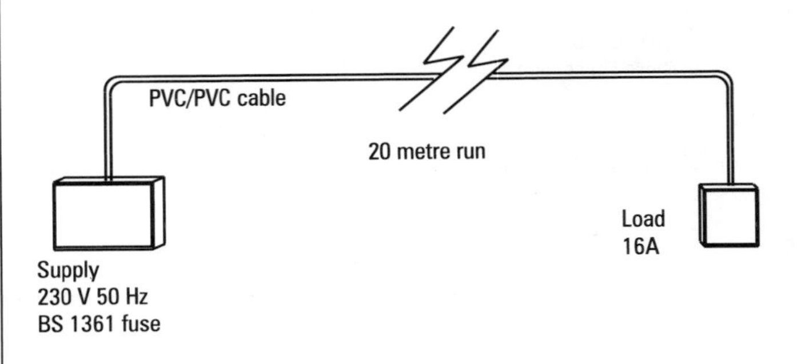

Figure 2.17

The first thing we shall do is to establish the installation method. We know how the cable is to be installed and so we can find the method reference number in BS 7671 Appendix 4, Table 4A. You can see that the table is divided into four columns. Column 1 gives a reference number to each method of installation and column 2 gives a written description of the method. The third column gives an illustration of the method that has been described and sometimes additional qualification of the requirements. All three columns are used to establish the way in which the cable is to be installed and once this is done we can read off the reference method from column 4.

For our example then we can see that the description in the first row of column 2 is the one that describes the way we are to install our cable so the reference method from column 4 is Method 1.

Now we can carry out the calculations to enable us to select conductor size.

$I_b = 16$ A and $I_n \geq I_b$ so from the tables we can determine that the size of BS 1361 fuse to be used is 20 A. As there are no factors to be applied then we can see that

$$I_t \geq \frac{20}{1 \times 1 \times 1 \times 1}$$

$$\geq 20 \text{ A}$$

Our next step is to select the cross sectional area of the conductor that we are to use and we do this with the aid of the tables in Appendix 4 of BS 7671. The cable size is selected on the basis of the current carrying capacity of the conductor under the stated conditions. This is where the method of installation is important because we use this to establish which column of the table to use.

We must select the appropriate table in Appendix 4 to find the correct size of conductor. Tables 4D1A to 4L4B contain the relevant information on current carrying capacity and voltage drop. If we look below the table heading, we find a description of the cable type, insulation and conductor material. Table 4D1A for example deals with single core PVC insulated non armoured cables with copper conductors. As we are not going to install single core cables, this table is not appropriate. We are to install PVC/PVC cables, and these will come under the heading of multicore cables. Table 4D2A covers the type we are using so this is the table we shall use.

The actual current carrying capacity of the cable is known as I_z and as you may imagine the value of I_z must be at least equal to the value of I_t for the circuit in question. So we can say that $I_z \geq I_t$.

Remember

I_z is the actual current carrying capacity of the cable

I_t is the tabulated current found in the tables

I_b is the design current of the circuit

I_n is the operating current of the fuse or circuit breaker

$I_z \geq I_t$ $\quad$ $I_t \geq I_n$ $\quad$ $I_n \geq I_b$

So our cable selection is made in the following stages:

1. Select the appropriate table
2. Select the vertical column that relates to the reference method that we are using
3. Select the column for the type of circuit, i.e. single or three-phase
4. Move down this column until a value that is equal or greater than I_t is reached
5. Move horizontally across the columns to column 1 and read off the cross sectional area of the cable

Now, for our example, we need a cable that has a current carrying capacity of at least 20 A and this will be 2.5 mm^2 cable. This is found using vertical column 6.

Once we have established the size of the cable we are to use we must also refer to the continuation of 4D2B which gives the voltage drop for cables in millivolts per ampere per metre. For a 2.5 mm^2 cable we get a value of 18 mV/A/m from column 3. We can now use this value along with the other information we have to determine the voltage drop along the length of the cable.

To find the volt drop we need to know the length of the cable, the amount of current it will carry and the volt drop per ampere per metre.
We put this into the formula as

$$\text{Volt drop} = \frac{\text{mV / A / m} \times I_b \times \text{length}}{1000}$$

This will give us a volt drop in volts.
In our example

$$\text{Volt drop} = \frac{18 \times 16 \times 20}{1000} = 5.76$$

Remember: the load current, I_b, is used to calculate voltage drop.

So now we know the actual voltage dropped along the length of the cable but does this comply with the maximum allowed? To find this out we must determine the maximum value. We will assume that for our calculation the voltage drop measured from the main intake position to any point on the installation must not exceed 4% of the nominal supply voltage. This is the maximum allowed in BS 7671.

Remember

The main intake position is the point at which the supply authority's supply cable enters the premises. Any distribution circuit voltage drop must be included in our calculations.

For our installation this is 230 V and so we have

$$\frac{4 \times 230}{100} = 9.2 \text{ V}$$

The actual volt drop must be equal to or less than the maximum volt drop allowed and in our example 5.76 V < 9.2 V so our circuit does comply with the volt drop constraints.

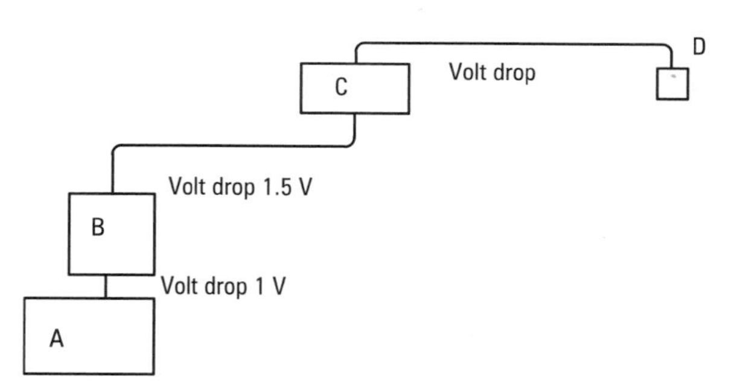

Figure 2.18 *If the maximum volt drop is 9.2 V then the volt drop from "C" to "D" must not be greater than* $9.2 - (1.5 + 1.0) = 6.7$ V

Remember

"C" is the correction factor to be applied:

C_a for ambient temperature
C_g for grouping
C_i for thermal insulation
C_f when semi-enclosed fuses are used.

Now let's have a go at a more complex installation.

A circuit is to supply a load of 2.75 kW with a power factor of 0.9 at 230 V 50 Hz. A PVC insulated and sheathed steel wire armoured cable is to be installed clipped to a perforated metal cable tray with a total cable length of 30 m from the distribution board. The distribution board is located at the supply intake position. For most of its length the cable is installed touching two other cables on an ambient temperature of 40 °C. If protection is provided by a BSEN 60269-1 (BS88) type fuse what is the minimum size of cable that can be used to comply with both current carrying capacity and voltage drop constraint if the equipment will not function at less than 224 volts?

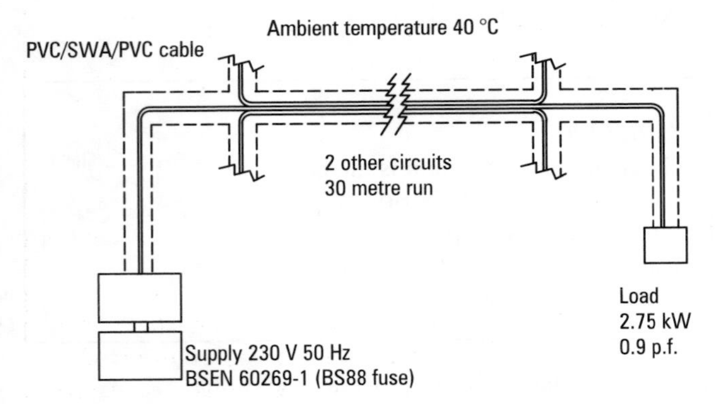

Figure 2.19

Solution: The reference method from Table 4A is method 11.

$$\text{power} = U \times I_b \times \text{power factor}$$

$$2750 = 230 \times I_b \times 0.9$$

$$I_b = \frac{2750}{(230 \times 0.9)}$$

$$= \frac{2750}{207}$$

$$= 13.285 \text{ A}$$

$I_b = 13.285\text{A} \therefore I_n \geq 13.285$

from table 41B1, for type BSEN 60269-1 (BS88) fuses, I_n = 16 A

$$I_b = \frac{I_n}{C_a \times C_g \times C_f \times C_i}$$

C_a = 0.87 from Table 4C1
C_g = 0.81 from Table 4B1
C_f = 1.0 as a BS3036 type is **not** being used.
C_i = 1.0 as cable is not surrounded by insulation

> *Remember*
> A cable installed touching two other cables has a grouping factor applied to the total number of cables so the factor is for three cables touching.
>
> 

$$I_t = \frac{16}{0.87 \times 0.81 \times 1 \times 1}$$

$$= \frac{16}{0.7047}$$

$$= 22.7\text{A}$$

Select cable size from 4D4A column 4.

1.5 mm^2 at 22 A is too small so we must use 2.5 mm^2 at a rating of 31 A.

$$I_t = 31 \text{ A}$$

mV/A/m from table 4D4B is 18 mV

Maximum volt drop = 230 – 224 = 6 V

Actual volt drop

$$= \text{mV/A/m} \times I_b \times \text{length}$$

$$= \frac{18 \times 13.285 \times 30}{1000}$$

$$= 7.17 \text{ V}$$

This does not comply as 7.17 > 6 V (the maximum allowed).

We must now go up in cable size in order to reduce the volt drop.

4 mm^2 has a mV/A/m of 11 mV

$$\therefore \text{volt drop} = \frac{11 \times 13.285 \times 30}{1000}$$

$$= 4.384 \text{ V}$$

As 4.384 V < 6 V maximum then this cable will be suitable.

So the smallest cable that we can use to ensure compliance with both current carrying capacity and voltage drop constraints is a 4 mm^2.

In this example we had to carry out the same calculation for voltage drop twice as the first selection did not comply. In some instances we may find that we do this calculation 3 or more times. We can avoid this by taking an alternative approach.

Once we know the values of I_b, length and the maximum permissible volt drop we can calculate the maximum value of mV/A/m that will comply with the volt drop requirements.

By using the formula

$$\text{Maximum volt drop} = \frac{\text{mV / A / m} \times I_b \times \text{length}}{1000}$$

we can see that

Maximum mV/A/m

$$= \frac{\text{maximum volt drop} \times 1000}{(I_b \times \text{length})}$$

If we use our example

$$\text{Maximum mV/A/m} = \frac{6 \times 1000}{13.285 \times 30}$$

$$= 15.05 \text{ mV/A/m}$$

Now when we select our cable size and go to the mV/A/m table we can see that as the maximum acceptable value would be 15.05 mV and 2.5 mm ^{2} has a value of 18 mV that it would not be suitable. We then continue down the column until we come to a value that is equal to or less than our maximum. By using this method we only have to carry out the calculation for volt drop once and then make a selection from the tables.

Try this

A circuit is to supply a load of 6 kW and unity power factor at 230 V 50 Hz a distance of 10m from the main intake position. As a degree of mechanical protection is required, a PVC insulated steel wire armoured PVC served cable is to be used and it is to be protected by a BSEN 60898 Type B MCB. The cable has to run through an area with an ambient temperature of 35 °C and is installed touching one other cable throughout its length. If the cable is to be clipped direct to a perforated steel cable tray determine the minimum size of the cable that can be used to comply with current carrying capacity and voltage drop constraints.

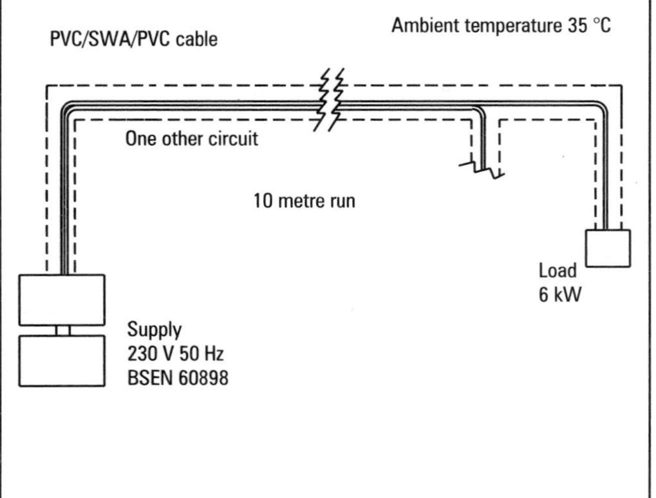

We have found that we may need to increase the cross sectional area of cables to comply with the volt drop constraint. In practice cable sizes are often dictated by volt drop rather than the current carrying capacity. It is therefore vital that this check is made and that selection is not done on current carrying capacity alone.

Try this

1. State four factors that will affect the volt drop over a length of cable.

2. If a cable carries a current of 15 A over a distance of 25 m and the maximum voltage drop allowed is 3 V calculate the maximum mV/A/m that would give compliance with the volt drop constraints.

3. A circuit supplies a load of 4 kW at a power factor of 0.95 over a distance of 25 m measured from a distribution board located at the main intake position. The supply is 230 V, 50 Hz and the circuit is protected by a BS1361 fuse. Determine the minimum size of light gauge MICC/PVC sheathed cable required to comply with current carrying and a maximum voltage drop of 6 V. The cable is installed clipped directly to a non metallic surface, exposed to the touch, and there are no factors to be applied.

4. A PVC insulated, LSF sheathed SWA cable has two conductors of 16 mm^2 copper. It is installed from the main incoming supply position to an isolator 50 m away clipped to a perforated steel cable tray. For most of its length it is run spaced from two other cables in an ambient temperature of 25 °C. The circuit is protected by a BSEN 60269-1 (BS88) type fuse. What is the maximum load that could be connected to the isolator to ensure that the voltage drop due to the cable does not exceed 3 V?

We have so far established the requirements for selecting the size of cables based on the live conductors. We now need to determine the method of sizing the protective conductors for our circuits to ensure compliance with the requirements for shock protection. To fully appreciate the requirements for shock protection we must be aware of the effects of an earth fault on

- the equipment supplied
- the whole of the installation
- the building structure

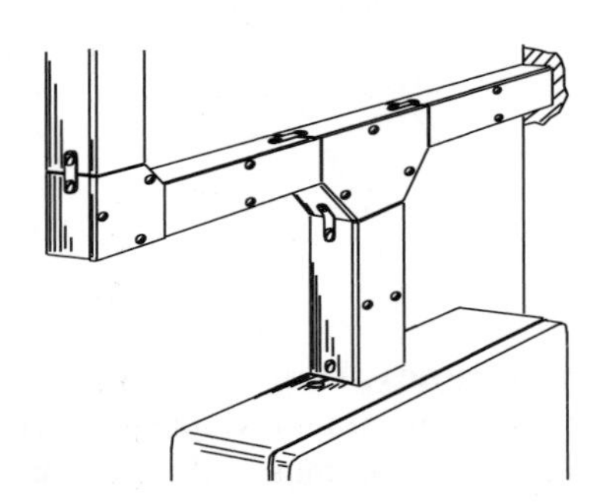

Figure 2.20 *All cpcs must be continuous throughout.*

Before we can begin there are a couple of terms that we must be familiar with. These define two components that may become part of the earth fault path and they are "Exposed Conductive Part" and "Extraneous Conductive Part". The definition of each of these terms can be found in BS 7671.

Exposed conductive part

An exposed conductive part is defined as:

"A conductive part of equipment which can be touched and which is not a live part but which may become live under fault conditions".

This means any exposed metal parts of the electrical installation. This is because any of these parts can become live in the event of an earth fault on any circuit within the installation as all the circuit protective conductors are connected to a common point. If the potential on any of these conductors rises above earth potential then all the parts connected to the common point will also rise above earth potential.

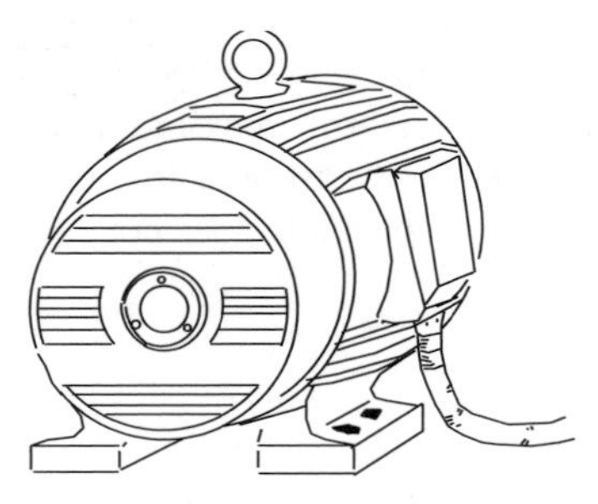

Figure 2.21 *The metal case of a motor is an exposed conductive part.*

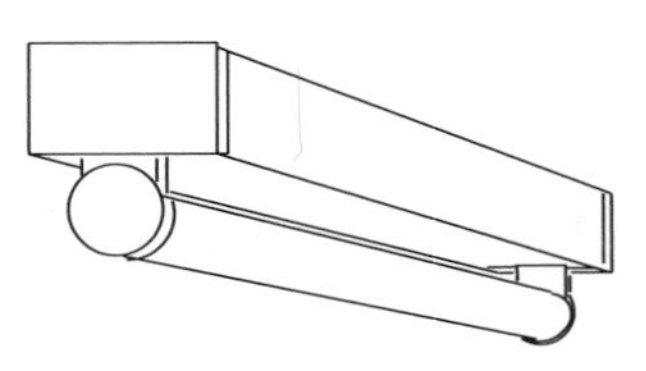

Figure 2.22 *The metal case of a fluorescent luminaire is an exposed conductive part.*

Extraneous conductive part

The definition of an extraneous conductive part is given as

"A conductive part liable to introduce a potential, generally earth potential, and **NOT** forming part of the electrical installation".

This means any structural steelwork, water pipes, gas pipes, drain pipes – in fact it can include any metalwork within the confines of the installation that is not a part of the electrical installation. We must satisfy ourselves as to the probability of the metalwork introducing a potential, including earth potential, to any point within the installation, this will establish whether there is a need to bond these parts to the common earth point.

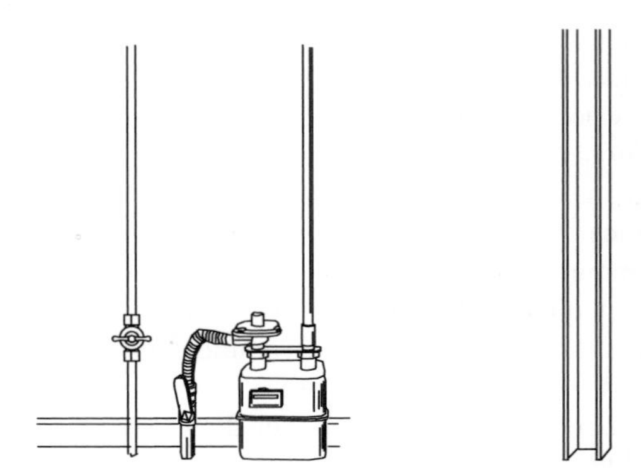

Figure 2.23 *Metal services and structural steel are extraneous conductive parts.*

Gas and water pipes will need to be protected against the effects of an earth fault current. Structural steel may or may not need this protection.

In order to protect both the installation and the consumer against the effects of an earth fault we must provide a safe return path for any earth fault currents. There are a number of ways to provide this return path and the method used will depend on the type of supply system to which the installation is connected. When a fault to earth occurs current flows around the earth fault loop (the safe return that we provide for faults). At this point it would be as well to remember that we actually require the highest possible current to flow. This will ensure the rapid operation of the protective device and the rapid disconnection of the circuit from the supply.

We will consider some aspects of the requirements for earthing and consider the types of systems later in this book. However, in order for us to complete the cable selection exercise we need to establish some of the requirements at this time, beginning with the

Earth fault path

The impedance of the earth fault path, known as Z_S, plays an important part in the system as it will regulate the amount of current that flows in the earth fault path.

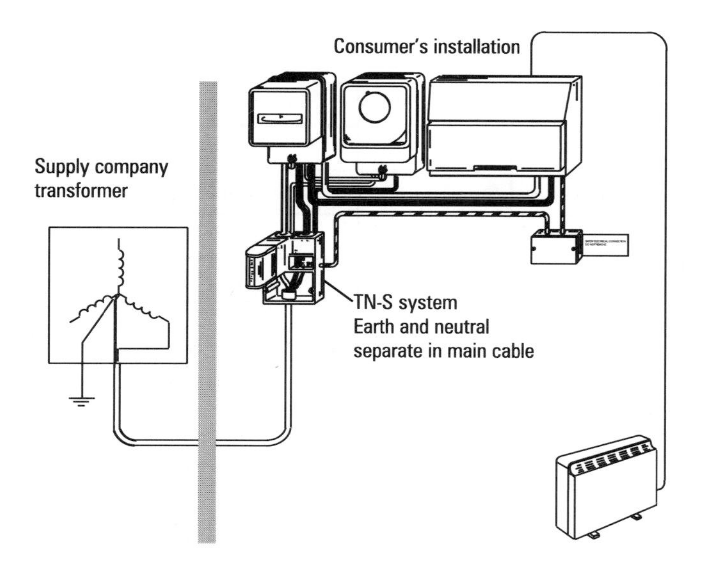

Figure 2.24 *TN-S system*
Earth faults are returned to the supply transformer through the metal sheath of the supply cables

If we look at the circuit diagram in Figure 2.25, we can see the case of the electric heater is connected to the circuit protective conductor. This is in turn connected to the consumer's earth terminal and then via the earthing conductor and the supply system to the star point of the transformer. All of which is best shown on the TN-S system for clarity.

All the parts of the circuit which are the consumer's responsibility, and connected to the consumer's earth terminal, are "exposed conductive parts". These are all part of the electrical installation and include conduit, trunking, circuit protective conductors and cases of appliances and equipment. Remember that in the case of a fault to earth all the "exposed conductive parts" of the installation become live for the period of time that it takes for the protective device to disconnect the circuit from the supply. The earth fault current will flow around the earth fault loop as shown in Figure 2.25.

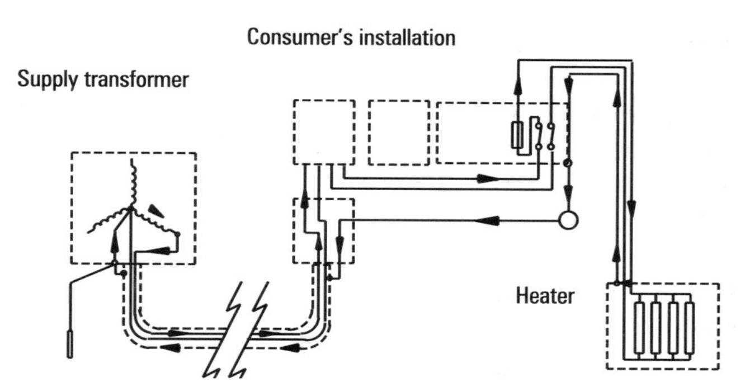

Figure 2.25 *Circuit diagram for TN-S system*
Arrows show the path that the current will take in the event of a fault.

This path will offer some resistance to the flow of current which will be dependent on the impedance of the conductors which go to make up the loop. As we can see, the loop comprises the transformer winding, the phase conductor of the supply and the consumer's circuit up to the fault. From the fault to the consumer's earth terminal is the circuit protective conductor and from the consumer's earth terminal, via the earthing conductor, back to the star point is dependent on the type of system. For our calculations we assume that the fault itself offers no resistance to the flow of current.

Figures 2.26 and 2.27 show the two other main types of system used in the public supply system in the UK.

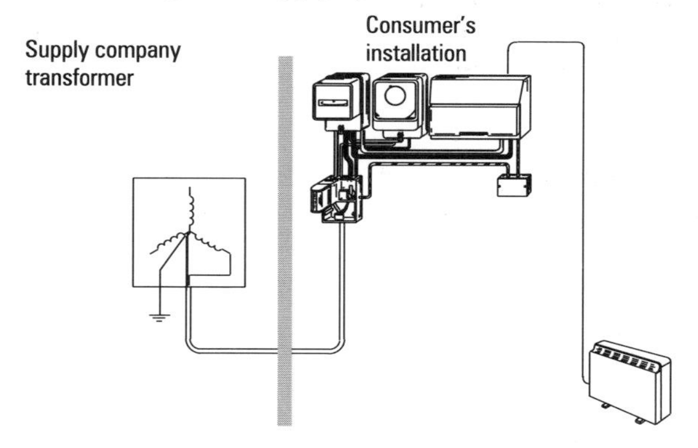

Figure 2.26 *TN-C-S system*
The supply cable has a combined earth and neutral conductor.

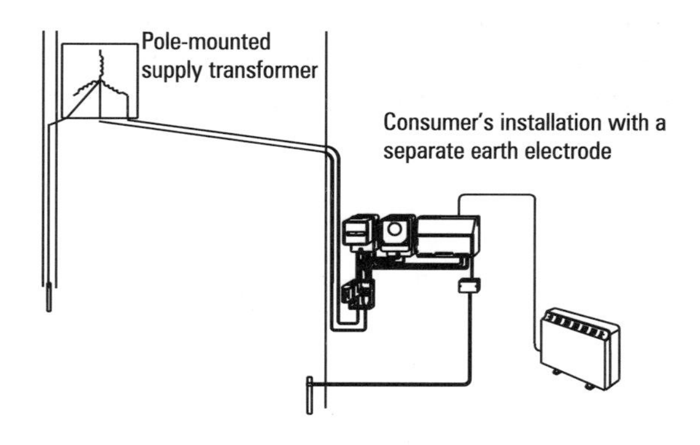

Figure 2.27 *Where the supply company is unable to provide an earth return to the transformer, the consumer requires a separate earth electrode.*
An RCD is the preferred type of protective device to ensure adequate protection.

In order for us to determine the current that will flow in the event of a fault we need to establish the impedance of this earth fault loop. If the installation is already installed then the earth fault loop can be measured using a line-earth loop impedance tester. If we are designing a system then we need to calculate the value of earth fault loop.

To do this we use the value of earth fault loop impedance of the supply system which we obtained from the supply authority, known as Z_e. To this we must add the impedance of the part of the loop that is made up by the circuit conductors. This will be the phase conductor up to the fault and the circuit protective conductor back to the consumer's earth terminal. These values are known as R_1 and R_2 respectively and are taken to the point on the circuit furthest from the supply to establish the worst case, that is, when the conductor impedance is at its maximum.

We can best see how this is done by using our example of the electric heater and calculating the earth fault loop impedance.

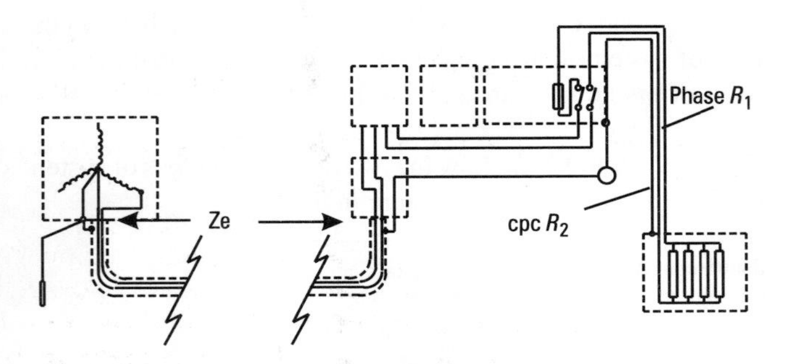

Figure 2.28

Conductor resistances

From earlier calculations we can assume we have established that the size of the live conductors we are using as 2.5 mm^2. The standard sizes of composite cables are supplied with the cpc one size smaller than the live conductors, so it would be a good idea to take this as a starting point for determining the value of the earth fault loop impedance. We know then that the size of the cpc is going to be 1.5 mm^2 and so we are ready to calculate the value of the earth fault loop impedance of the consumer's part of the system (R_1 and R_2).

To carry out the calculation we need to know the following details

- the length of the circuit conductors
- the cross sectional area of the phase conductors
- the cross sectional area of the circuit protective conductor (this may be the same as that of the phase conductor but not always)
- the impedance of the phase and protective conductor per metre

This last detail we can get from tables, so let's take a look at the section of these as shown in Table G1 in Guidance Note 1.

Try this

Complete the tables below using tables in IEE Guidance Note 1.

Table 2.6

Values of resistance/metre for copper conductors at 20 °C in milliohms/metre.

Cross-sectional area mm^2		Resistance milliohms/metre
Phase conductor	Protective conductor	Plain copper
1	–	
1.5	–	
1.5	1.5	
2.5	–	
2.5	1.5	
2.5	2.5	
4	–	
4	2.5	
6	2.5	
6	4	

To allow for temperature rise under fault conditions the following correction factors must be applied.

Table 2.7

Insulation material	70 °C PVC
Multiplier	

The important thing to remember is that the values given are in MILLIOHMS per metre.

For our calculation we require the resistance of a 2.5 mm^2 phase conductor with a 1.5 mm^2 protective conductor. From Table 2.6 this is 19.51 mΩ/m. This is not the end of the calculation though as we need to know the resistance under fault conditions and this could mean the temperature rising in the conductors and increasing their resistance. To allow for this a factor of 1.2 must be applied to our total value.

We now have all the details that we need to calculate the value of the earth fault path within the consumer's installation. This will be R_1 and R_2.

Assume	the length of run	= 20 m
	resistance from Table 2.6	= 19.51 mΩ/m
	multiplier from Table 2.7	= 1.2

Remember: Always use the multiplier from Table 2.7.

The total value will be calculated using the formula:

$$\frac{\text{resistance}}{(R_1 + R_2)} = \frac{\text{m}\Omega/\text{m} \times \text{length} \times \text{multiplier}}{1000}$$

$$\text{resistance} = \frac{19.51 \times 20 \times 1.2}{1000}$$

$$= 0.46824\ \Omega$$

Try this

Determine the value of R_1 and R_2 from Table 2.6 for

1. 1.5 mm^2 phase conductor with 1.5 mm^2 cpc

2. 2.5 mm^2 phase conductor with 2.5 mm^2 cpc

3. 6.00 mm^2 phase conductor with 2.5 mm^2 cpc

Earth fault loop impedance value

To establish the total value of earth fault loop impedance we must add to this the earth fault loop impedance of the supply, Z_e. We shall assume the supply company has quoted a value of 0.35 Ω for this, so we get a total of

$$0.35 + 0.46824 = 0.81824\ \Omega$$

The earth fault loop impedance is found using the formula

$$Z_s = Z_e + (R_1 + R_2)$$

So what is the significance of the value of Z_s?

If we refer to tables 41B1, 41B2 and 41D in BS 7671 we find the types and ratings of protective devices listed. Below each rating is given a maximum value for Z_s for the device.

Tables 41B1 and 41B2 give the maximum values of Z_s for devices supplying circuits incorporating socket outlets. Tables 41D and 41B2 give maximum values for circuits supplying fixed equipment.

We can see that the maximum values of Z_s are different for each table. This is because the disconnection time required for a circuit supplying socket outlets is given as within 0.4 seconds. Circuits supplying fixed equipment must disconnect within 5.0 seconds. The value of Z_s for fixed equipment can therefore be higher than that for socket outlets as a longer period for disconnection is allowed, therefore a lower current can flow.

We will assume that on our circuit supplying the electric heater we are using a 15 A BS1361 type fuse to supply fixed equipment. We use Table 41D to check for compliance of the circuit. The maximum value of Z_s is given in the table as 5.22 Ω for a 15 A BS1361 type fuse. As our value is 0.81824 Ω, this circuit does comply.

If we apply a simple Ohm's law calculation to our circuit then we can find the current that will flow in the event of a fault to earth. This is known as the prospective earth fault current I_f and is found by using the formula

$$I_f = \frac{U_o}{Z_s}$$

where:

I_f is the prospective fault current

U_o is the nominal phase voltage to earth and

Z_s is the earth fault loop impedance

$$\text{so } I_f = \frac{230}{0.81824} = 281\ \text{A}$$

Now that we know the value of the prospective fault current, we can further check the compliance of the circuit by using the tables in Appendix 3 of BS 7671. These tables are the simplified time/current curves for the various types of protective devices.

The horizontal axis is the prospective fault current and the vertical axis is the time. The divisions along the axis are logarithmic and, as we can see, the divisions are not equal in size. Figure 2.29 shows a section of the X (horizontal) axis of the graph. We can see that the origin of the axis, the extreme left-hand end, has a value of 1 NOT 0 and the divisions go up in units of one until we reach a value of 10. The divisions then go up in units of 10 until a value of 100 is reached at which point the divisions go up in units of 100 and so on. In this way a considerable current range can be covered in a relatively small space.

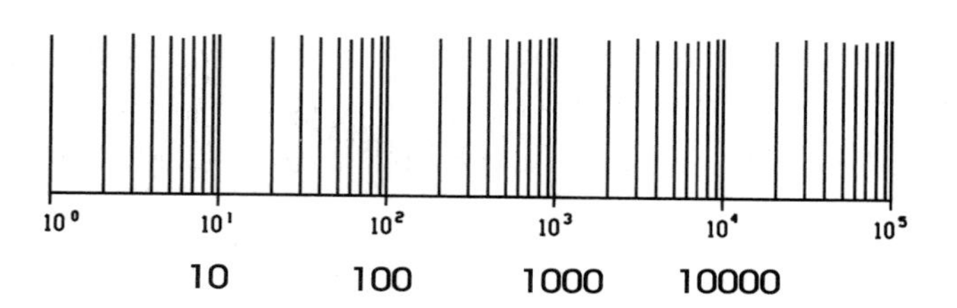

Figure 2.29

Figure 2.30 shows a section of the X axis where we've had to subdivide one of the sections. We must remember that subdivisions of scale sections must be done in the same ratio as the main scale. This means that the value 250 A is NOT going to be halfway between 200 A and 300 A. It occurs at a point approximately 0.66 of the way between these two as shown on Figure 2.30.

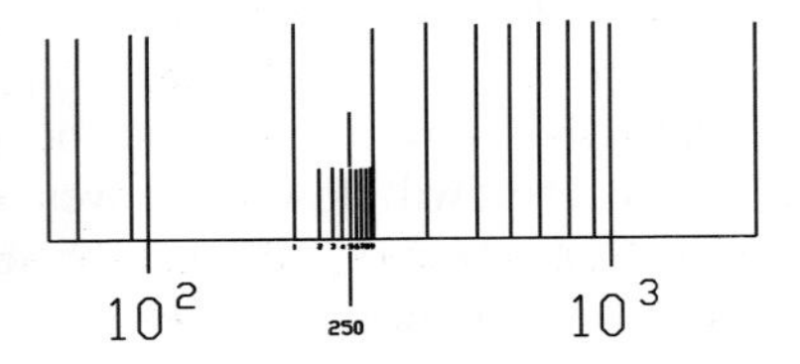

Figure 2.30 *The value 250 is not halfway between 200 and 300 but nearer two-thirds.*

The vertical Y axis is shown in Figure 2.31 and again we can see that the origin of the axis is NOT 0. For ease of location the 1 second position is highlighted on the axis in Figure 2.31. By working back down the axis we can see that the origin of the axis is in fact 0.01 seconds. Like the X axis the divisions are logarithmic and so the same proportional division of the scale must be used.

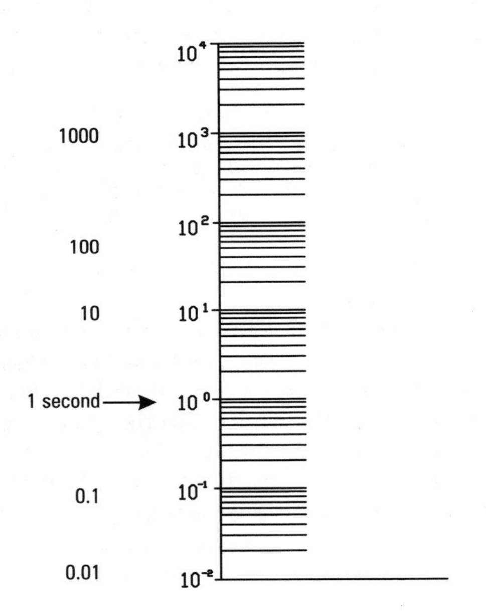

Figure 2.31

Now to further check the compliance of our circuit we can use the value of I_f and the appropriate time/current curve to establish the disconnection time for the device. To do this we use the appropriate table for BS1361 fuses in Appendix 3, BS 7671or a manufacturer's details. Figure 2.32 shows the appropriate section of the graph.

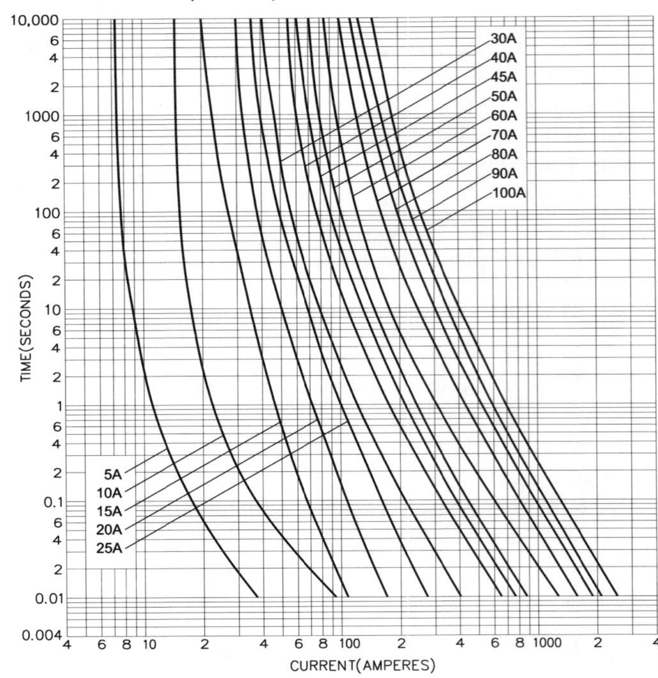

Figure 2.32 *Time/current characteristics for fuses to BS 1361.*
Reproduced with kind permission of Bussmann Division, Cooper (U.K.) Ltd.

Try this

Mark the positions on the scale of the following currents: 6 A, 550 A, 65 A, 1250 A, 275 A

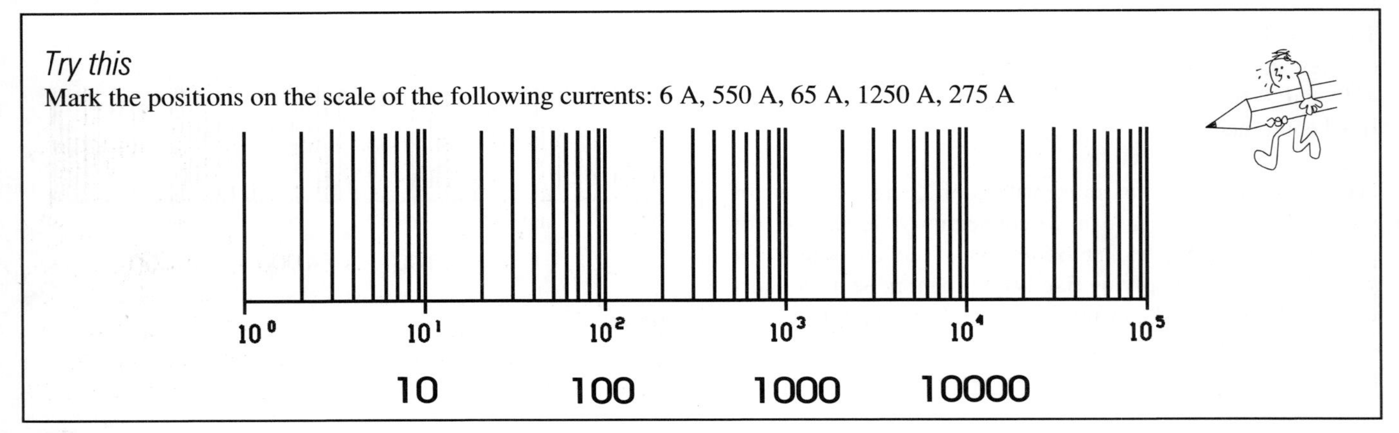

We move along the X axis until we reach the value of I_f for our circuit (281A), we then move vertically up from this point until we bisect the curve for the 15 A device. A rule will be of some help when carrying out this exercise. We then move horizontally across from this point to the Y axis and read off the time. As the circuit we are considering supplies fixed equipment the disconnection time must be no more than 5 seconds. We can see that a 15 A BS1361 type fuse requires a current flow of approximately 67 A to achieve disconnection within 0.1 sec. The value we get with a current of 281 A is in fact less than 0.1 seconds so our circuit complies with requirements, but we expected this as the maximum value of Z_s already indicated this to be the case.

TIP

On your own copy of the time/current curve use a highlighter to mark across the curves at the 5 s and 0.4 s positions for ease of reference.

Try this

Mark the positions of the following times on the axis.

1 s, 0.4 s, 5 s, 0.75 s, 1.4 s

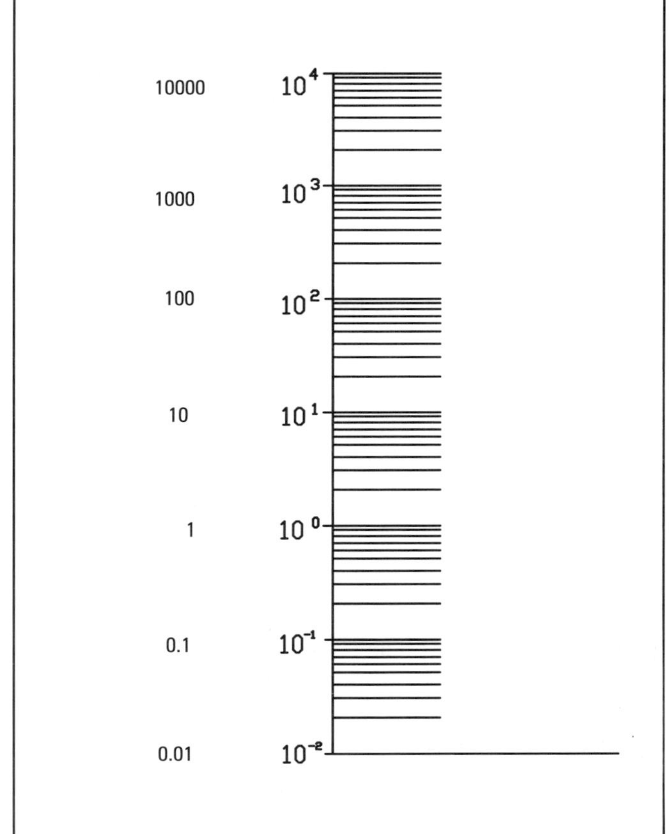

So we have found the size of conductors needed to give compliance with requirements for shock protection. Whilst doing this we found that a high current flows in the event of an earth fault under the correct conditions. To provide shock protection this current must be carried by the circuit protective conductor for the time that it takes for the protective device to operate and disconnect the circuit from the supply. As the circuit protective conductor is generally a smaller cross sectional area than the live conductors, and the current that it carries may be quite high, a great deal of heat will be generated whilst the current flows.

We must make sure that, whilst the fault current is flowing, the heat produced will not cause damage to the conductors or the insulation and material surrounding them. If sufficient current flows for long enough the heat produced could be such that the insulation catches fire. Once this happens disconnection of the supply will not extinguish the flames.

Obviously this situation cannot be allowed to occur. To prevent it we must ensure that the circuit protective conductor is large enough to carry the fault current for the period of time needed for the device to disconnect the supply without excessive heat being produced.

The thermal constraint placed on the circuit by BS 7671 ensures that the circuit protective conductor of the circuit is large enough to carry the earth fault current without a detrimental effect on the conductor, the insulation or the installation.

In order to find the minimum cross sectional area of the circuit protective conductor we require the following information

- the prospective earth fault current "I_f"
- the time taken for the protective device to operate with this fault current "*t*"
- the constant *k* which is related to the type of circuit protective conductor and its method of installation

The way in which these are related to the minimum size of the circuit protective conductor is by a formula known as the adiabatic equation.

The minimum cross sectional area of the conductor is known as "S" and the formula we use is

$$S = \frac{\sqrt{I^2 \times t}}{k}$$

(The value of I_f is used for this calculation.)

It is important that the calculation is carried out in the right order to give the correct answer.

Let's consider the circuit supplying the electric heater to see if it complies with the requirements for thermal constraints. We know that the prospective earth fault current is 281A. We also found that disconnection time will be less than 0.01 of a second. Now we need to find the value of the constant *k* for our circuit. The values for *k* are given in tables 54B to 54F in BS

7671. At the head of each table is the description of the method of installation of the circuit protective conductor. For our type of cable and the method of installation we find the value of k from table 54C to be 115.

Try this

Complete the following table using Table 54C from BS 7671.

Table 2.8 *Values of k for protective conductor as a core in a cable or bunched with cables*

Material of conductor	Insulation material		
	70 °C PVC	85 °C rubber	90 °C thermosetting
Copper			
Aluminium			

If we now put these values into the formula we have

$$S = \frac{\sqrt{281^2 \times 0.1}}{115}$$

We must carry out the calculation in the correct sequence to give the correct solution

stage one: Square the value of I_f, i.e. $I_f \times I_f$
stage two: Multiply the result by the time t second
stage three: Take the square root of the result
stage four: divide the answer by the value of k

In our particular case this would be

$281 \times 281 = 78961$

$7896 \times 0.1 = 7896.1$

$\sqrt{7896.1} = 88.86$

$\frac{88.86}{115} = 0.773\ \text{mm}^2$

As the minimum size of cpc to give compliance is 0.773 mm^2 and we have installed a cpc of 1.5 mm^2 we can see that circuit fulfils the requirements for thermal constraints and so our circuit complies.

If the size of cpc installed proves to be insufficient then a larger cross-sectional area conductor must be used. This will, of course, have an effect on the value of the earth fault loop impedance as a larger cpc will reduce the impedance and a higher current will flow. This in turn will reduce the disconnection time.

We could calculate the minimum requirements based on the known criteria that apply to our circuit. For example we know that for a socket outlet circuit the maximum disconnection time is 0.4 seconds. If we know the type and rating of the protective device we can establish the minimum value of I_f to give disconnection in 0.4 seconds. This data and the value of k for the type of cpc and its method of installation will allow us to establish the minimum size of cpc to give compliance under the worst conditions.

To do this we must use the minimum value of I_f to give the disconnection time required, the maximum disconnection time and k for the type and method of cpc installation. If we carry out this one calculation we can establish the minimum size of cpc to comply with the absolute worst conditions and we can ensure that the size selected will be within a usable range. This can be particularly useful when designing circuits for installation in conduit and trunking where the size of cpc can be varied with ease. It will also ensure that a great deal of time is not wasted carrying out repeated calculations to establish an acceptable size.

We shall consider the earthing requirements at a later point in this book.

Try this

1. A circuit supplies an item of fixed equipment from a 230 V 50 Hz supply. If the circuit protective conductor is 2.5 mm^2, the value of k is 115 and the disconnection time is to be 0.4 seconds what is the maximum value of earth fault current that can flow for compliance with the thermal constraint requirement?

2. Determine the minimum cross sectional area of protective conductor for a 230 V single-phase circuit which has the following?
   i) a value for Z_e of 0.3 Ω
   ii) a value for $R_1 + R_2$ of 0.7 Ω
   iii) a circuit protective device of 30 A to BS1361
   iv) a k factor of 143

3. (a) What is the minimum fault current that needs to flow through a 100 A fuse to BS 88 parts 2 & 6 if it is to operate in 0.4 seconds?
   (b) If the protection device in (a) was replaced with a fuse to BS 3036 when the same fault current was flowing how long would it take the new fuse to operate?

It is important to remember that, whilst we have considered some final circuit requirements for our cable selection, the same considerations must be given to the selection of cables for high voltage distribution cables. We need to determine maximum demand, consider diversity and then calculate the size of conductor and composite cables as we have here.

The maximum demand/load applied in such instances is generally based upon figures established for the final use and requirements of the installation to be served. This will then dictate the capacity of the substation transformer, which in turn will determine the loading for the high voltage system.

It is important to remember that one of the principal advantages of transmission and distribution at high voltage is the reduction in current for the same load. Therefore when calculating the size of high voltage distribution cables the conductor sizes are likely to be smaller than those used for the low voltage distribution circuits. This will be reflected through each transformer in the system. Whilst the conductor sizes are not as large the insulation requirements increase and the overall cable size may be larger. This is due to the extent of insulation required in the high voltage cables to prevent leakage at the voltages used. This is one of the reasons why the transmission of electrical energy is undertaken by overhead lines rather than buried cables.

When applying diversity to main and distribution cables we need to consider the connected load in accordance with BS 7671 and apply the appropriate diversity to each section of the load. Guidance Note 1, Selection and Erection published by the IEE gives suggested diversity for circuits based upon type of circuit and type of installation.

If there are a variety of different types of circuits then each type will need to have diversity applied to their group. Having determined the load for each group of circuits after diversity we can then determine the total load for the main or distribution cable.

Remember

Diversity is the ratio between maximum demand and connected load:

$$\text{Diversity} = \frac{\text{Maximum demand}}{\text{Connected load}}$$

where maximum demand is the maximum anticipated demand and connected load is the sum of all the electrical loads.

It is important to allow some capacity, over that currently required, for future additions to the installation. Once a consumer has a comprehensive distribution system installed, it is expected to be suitable for use for a number of years. The need to upgrade transformers, cables, distribution boards and the like at the first hint of change would indicate a poor design for an installation which may be expected to last up to 30 years or even longer.

It is not uncommon to allow a percentage of the known maximum demand for future extensions. Whilst this will depend on the type of installation and the likely developments in growth and technology, the usual minimum is 10%.

Many of the requirements for LV installations are applicable to the 11 and 33 kV installations. We must take care to balance loads as best we can across the three phases and give the same consideration to the discrimination of the protective devices.

In the case of supplies to transformers at 11 and 33 kV we can, at this stage, consider these as balanced loads for the purpose of the selection of the cables. However a load connected to our transformer that is not balanced will be reflected to the supply side of the transformer and hence back through the system.

Imagine a single generator unit supplying the transmission and distribution system and an unbalanced load connected to our final 11 kV: 400 /230 V transformer. The effect will be that the transformers connected in the system will become unbalanced as the imbalance on the output will be imposed on the input. At the final analysis this would result in the generator becoming unbalanced and cause considerable stress on the machine and inefficiency in operation.

The Public Electricity Suppliers have a number of generators but they have millions of consumers, which means that it is difficult to balance the loads evenly throughout the system. This is one of the reasons why the supply companies insist that any installation connected to their system is balanced as closely as possible across the three phases. We can see this requirement practised by the supply companies in the way they connect domestic consumers on different phases, as shown in Figure 2.33. This is often calculated across a new development to ensure that, whatever connection arrangements are used the result is that the development is fairly equally distributed.

Try this

Locate a three-phase distribution board which supplies a mixture of three phase and single phase circuits. Note the final circuit ratings for each circuit and determine the maximum demand that could be required for each phase. Determine from the results how well the load on that distribution board has been balanced across the three phases.

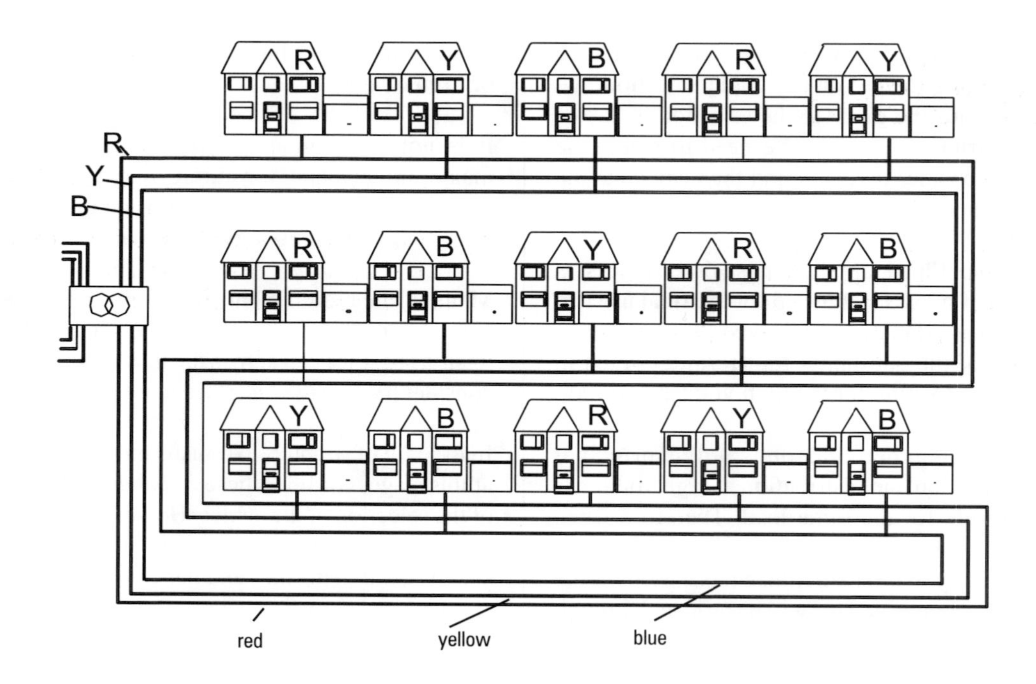

Figure 2.33

Try this

1. State the two methods of determining the maximum demand for a proposed installation in the early stages of design, giving an example of the type of installation for which each would be used.

2. What is meant by the "design current" of a circuit and what symbol is given to denote it?

3. List ALL the information that you would need to obtain from the supply company to enable you to select a suitable system of wiring and size the cables required for a proposed installation.

4. If a single phase load of 3.5 kW is to be supplied from a 220 V 50 Hz supply and has a power factor of 0.95 what will the circuit design current be?

5. Calculate the power factor for a 25 kW balanced three phase load if it is supplied at 450 V 50 Hz and has a line current of 45 A.

Exercises

1. It has been determined that a fault current of 10 kA is likely to occur at a point on a 33 kV, three-phase distribution system. Calculate
   (a) the symmetrical fault level that this will produce and
   (b) the asymmetrical fault current at $t = 0$ seconds.

2. State 3 advantages and 3 disadvantages for
   (a) a radial distribution system
   (b) a ring distribution system

3. A small business premises consists of the following single phase loads:

   26 lighting circuits rated at 6 A
   12 ring circuits at 32 A
   4 instantaneous water heaters at 8.5 kW 230 V
   28 motor circuits at 16 A

   Determine the maximum demand after the application of diversity in accordance with Guidance Note 1 and suggest a suitable method of supply for the building in single or three-phase with suggested capacity to allow for expansion.

4. Draw a simple line diagram of the supply network showing typical voltages at each stage.

3

Distribution Systems and Equipment

Revision questions:

1. List one advantage and one disadvantage for both the radial and the ring distribution system for High Voltage supplies.

2. List three principal advantages for large consumers in purchasing electricity at high voltage.

3. What is the checklist sequence for determining the size of cables and protective devices?

4. Why is it important to try to balance the single phase loads over the three phases?

On completion of this chapter you should be able to:

- describe the construction of cables used for distribution systems
- describe the construction of busbar and cable type distribution systems
- identify the advantages of modern materials and techniques related to installation
- identify the advantages of materials used in cable construction
- select suitable systems for given criteria
- list the considerations and requirements to be taken into account when selecting routes for distribution systems

In this chapter we shall consider some of the distribution systems available, in particular, cables and their applications and bus bar systems. This will include cable selection and installation including accessories, joints and terminations. We will also look at some of the many types of cables available, their typical uses and identify some of the factors we should consider when selecting them for distribution systems.

Before we begin let's take a few moments to remind ourselves of the basic components of cables and their general construction. All types of electric cable are made up of a low resistance conductor, or conductors, to carry the current, and insulation to isolate the conductors from each other and their surroundings.

The basic components of low voltage cables will usually be conductors and insulation but as the voltage increases, the construction becomes more complex as we shall see later in this chapter. Other components in an electric cable may include a sheath to provide protection and keep out moisture and armouring for mechanical protection.

Where the cable we select is available with BASEC (British Approvals Service for Electric Cables) certification, use of the BASEC cable will ensure that the cable has been manufactured and tested in accordance with the appropriate British or International Standards.

Before we can select the type and size of cable for any particular application, we must consider a number of factors which affect our final choice. These will include:

- voltage and frequency of supply
- normal full load current
- cable route, its length and method of installation
- proximity to other cables
- any correction factors required
- short circuit rating and type of system protection
- voltage drop
- suitability for external influences
- ambient temperature
- overall diameter, flexibility and minimum bending radius
- type of termination and termination compartments/enclosures

As you will have noticed the list of considerations is very similar to that applied to cable selection for any circuit, of course we do need to consider different factors dependent upon the method of installation. Underground cables, for example, may have a higher current carrying capacity as a result of the low ambient temperature of the soil in which they are buried. Current carrying capacities for buried cables are not listed in BS7671 but may be obtained by reference to manufacturers' data.

We can make a cable selection based upon the above factors. However, there are some additional considerations that we should take into account. Let's first look at the most common types of conductors and compare them.

Copper or aluminium?

Solid or stranded copper conductors are the most common in electrical installations, the main exception being for distribution circuits. The principal alternative conductor material is aluminium and this is generally a solid conductor irrespective of its size. Aluminium has a higher resistivity than copper and so a larger cross-section is required to enable the same current to be carried. Whilst aluminium cables are generally cheaper than copper cables the cost of jointing and terminating tends to be higher.

More space is normally required to terminate solid aluminium conductors and so more space will be required in distribution equipment, termination chambers and accessories. A further possible disadvantage of using aluminium cables is that their minimum bending radius is greater than that of equivalent copper cables.

Aluminium does have an electrolytic reaction when it comes into contact with brass. This means that where switchgear has brass terminations special adaptor lugs or pin terminals are required to ensure the connections do not corrode. These may be special alloy lugs or alloy to copper pin terminations for tunnel terminals.

Whilst there may be some disadvantages to using aluminium conductors there are some advantages that should be considered. The principal advantages are savings in initial cost and cable weight. The latter may be a major consideration during construction and the effect on the structure and support systems, to such an extent that it outweighs any disadvantages.

Multicore cables for power distribution will normally have shaped conductors.

> *Remember*
>
> BS7671 requires all cables used for low voltage to comply with the appropriate British or Harmonised Standard.
>
> 
>
> The use of BASEC approved cables is one way to achieve this. Alternatively we should ask the main manufacturer for written confirmation of compliance with the required standards.

You should be familiar with general purpose PVC insulated, sheathed and armoured cables and so we shall now take a look at the variety of types and construction of cables which are available. We shall also consider typical applications for each type.

We should be familiar with general purpose PVC cables and so we shall not consider them in any great detail here, other than to illustrate the typical constructions shown in Figures 3.1A and 3.1B. They have been the mainstay of electrical installation cabling for many years. However its thermoplasticity (PVC becomes soft when warmed and hardens when cooled throughout its working life) is the prime limitation. It is this factor that affects the current carrying capacity, overload and short circuit rating of cables which use PVC as the insulation material.

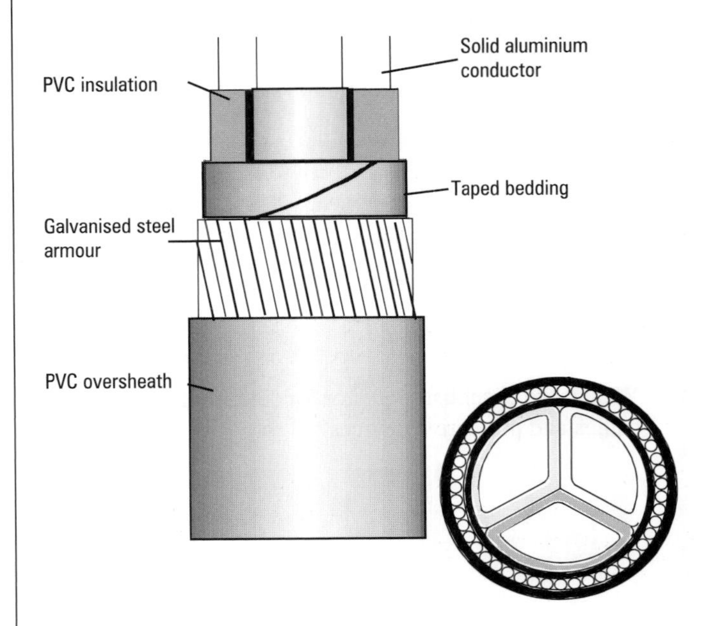

Figure 3.1a *Three-phase 600/1000 V PVC insulated SWA cable with PVC oversheath*

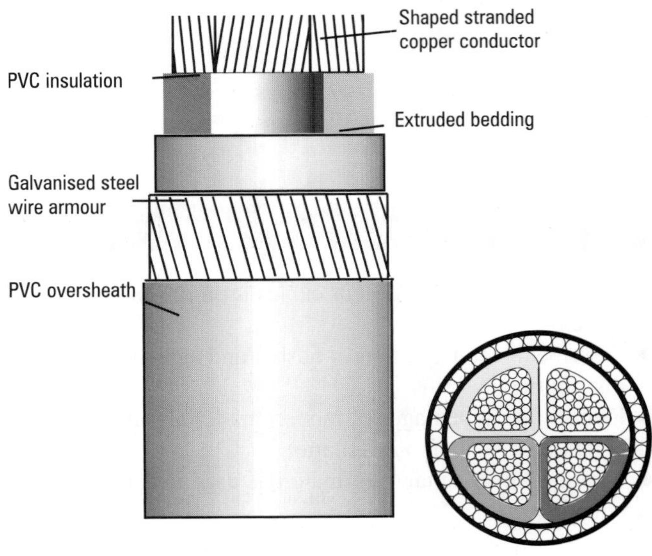

Figure 3.1b *Three-phase 4-core 600/1000 V PVC insulated SWA cable with PVC*

Cross-linked cables

To overcome the limitations of PVC we can use XLPE (cross-linked polyethylene) and EPR (ethylene-propylene rubber) thermosetting insulated cables. Although XLPE and EPR cables are more expensive than their PVC equivalents, the additional cost may be offset to some degree. This is because we may be able to use a cable with a conductor of smaller cross sectional area for the same current carrying capacity due to the characteristics of the insulation.

Because these are thermosetting materials, they are hardened fully after one application of heat, and do not vary during their working life. XLPE and EPR insulation offers additional protection as they are much tougher insulations and have the ability to withstand much higher temperatures. The more rugged physical properties of the cross-linked cable allows the insulation thickness to be reduced and so the overall size of the cable may be smaller.

Cables constructed with these cross-linked insulation materials have greater flexibility and a higher operating temperature limit than say, paper or PVC insulated cables, and in suitable circumstances may be used for voltages up to 33 kV.

Of the two cross-linked cable types, those with EPR insulation are more expensive to purchase than those with XLPE insulation, therefore EPR cables should only be used where the conditions require the cables to be flexible.

The maximum conductor operating temperature of cross-linked insulated cables is 90 °C compared to 70 °C for PVC insulated cables. It follows therefore that the surface temperature of the cross linked cable will also be higher when the cable is fully loaded at around 40 °C. We must take care when we select the route of XLPE and EPR cables to avoid locations where they may be easily touched etc. In most cases our cables will not be fully loaded so the conductor temperature is likely to be lower than the maximum 90 °C. This being the case the surface temperature is also likely to be reduced.

The construction of these cables is generally the same as for the PVC/PVC cable – only the insulation material is changed.

Remember

The use of cross linked cables may offer some advantages in cost, size and flexibility but the increased operating temperature must be considered. This may affect the selection of equipment and cable routes when such cables are selected.

Waveconal cables

Waveconal construction refers to a particular construction for Combined Neutral and Earth distribution cables, generally manufactured for the supply industry to use on TN-C-S systems. The phase conductors are generally solid aluminium insulated with XLPE or EPR covering. The neutral conductor is an aluminium wire which is concentrically applied in a waveform pattern. This configuration provides advantages during both the construction and installation of these cables.

The combined neutral and earth conductor is completely contained in a bedding of unvulcanised rubber which protects it against corrosion. This is then given an overall layer of extruded PVC which forms the outer sheath of the cable.

This type of cable is typically used for main distribution up to 1000 V although higher ratings are available. One of the features of the waveform application of the combined neutral earth conductor is that it allows access through to the phase conductors without the need for cutting the CNE conductor and so it is particularly useful for underground cables with service joints. A typical waveconal construction is shown in Figure 3.2.

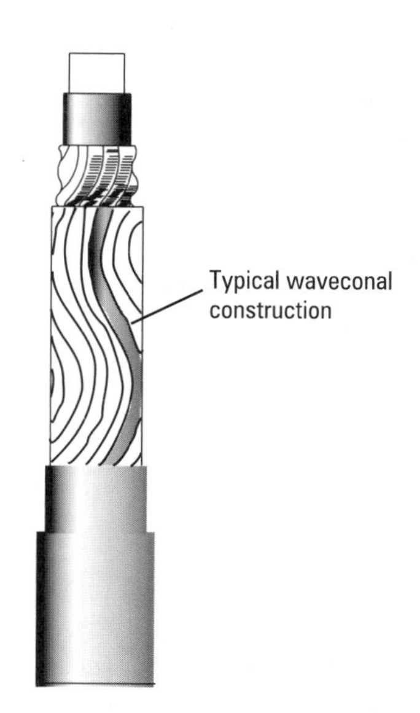

Figure 3.2 Waveconal single core mains cable

Split concentric cables

A typical use for this cable is for single phase supplies and the reason for this is fairly obvious from the construction. A single circular phase conductor, usually insulated with red PVC or XLPE, forms the central core of the cable. A concentric layer comprising in part of bare conductor earth and black covered neutral conductors are bound onto the phase conductor. It is common to find two PVC "fillers" separating the neutral and earth conductors, these have no other purpose than to make up space and produce a circular cable. The covering on the neutral conductors is not generally regarded as insulation.

A non hygroscopic tape is then applied around the outer layer of conductors and an overall sheath, usually of black PVC, is applied to complete the construction, as shown in Figure 3.3. This type of cable can be obtained as a three-phase concentric but this is far less common. These cables have a similar construction with the centre section having three shaped conductors.

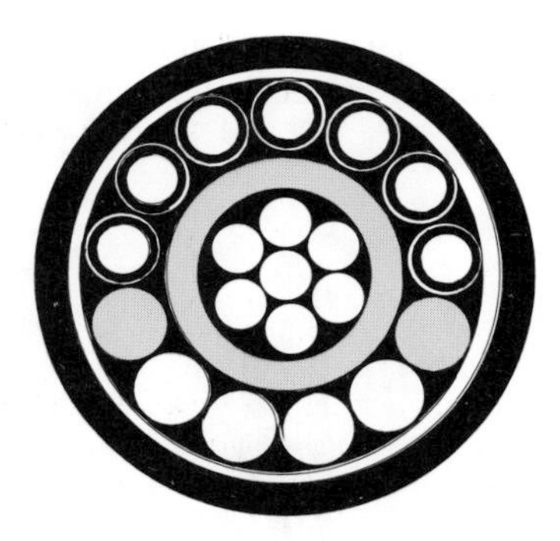

Figure 3.3 PVC insulated split concentric cable

Concentric cable is also available as a concentric CNE cable as shown in Figure 3.4 for use on TN-C-S supplies.

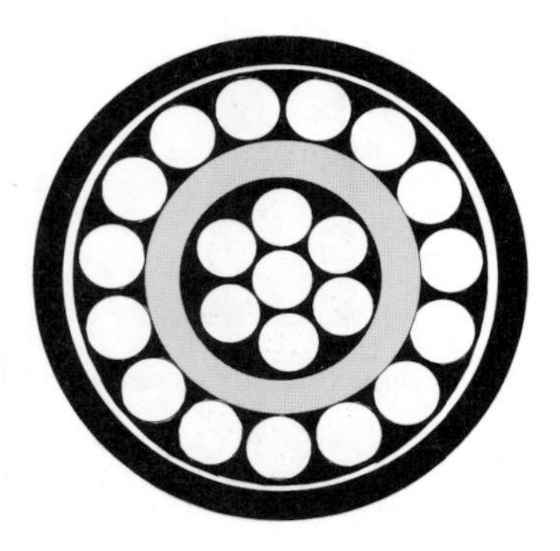

Figure 3.4 PVC concentric neutral cable

One of the advantages of this construction is the relatively small overall diameter of the cable in relation to the conductor size, typically a 25 mm^2 c.s.a. phase conductor would result in a cable with an overall diameter of 18 mm. The most common use for this type of cable is to provide services to dwellings or equipment supplied directly from the Public Electricity Supplier's main cable or feeder pillar.

PILCSTA and PILCSWA

Paper Insulated Lead Covered Steel Tape or Wire Armoured

Paper insulated cables have been the principal cables used for the supply and distribution of electricity for many years, by both the Public Electricity Suppliers and private suppliers. The principle of construction has remained almost unchanged throughout and typical examples are shown in Figures 3.5 and 3.6.

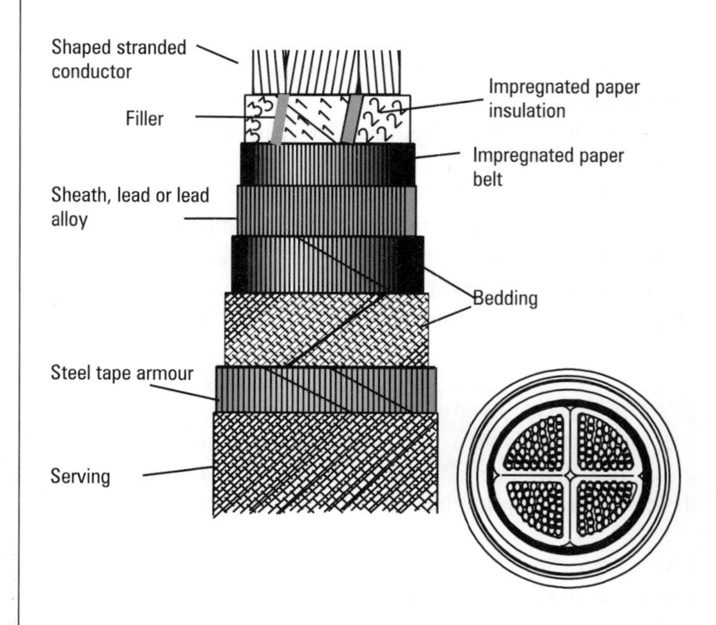

Figure 3.5 PILCSTA cable

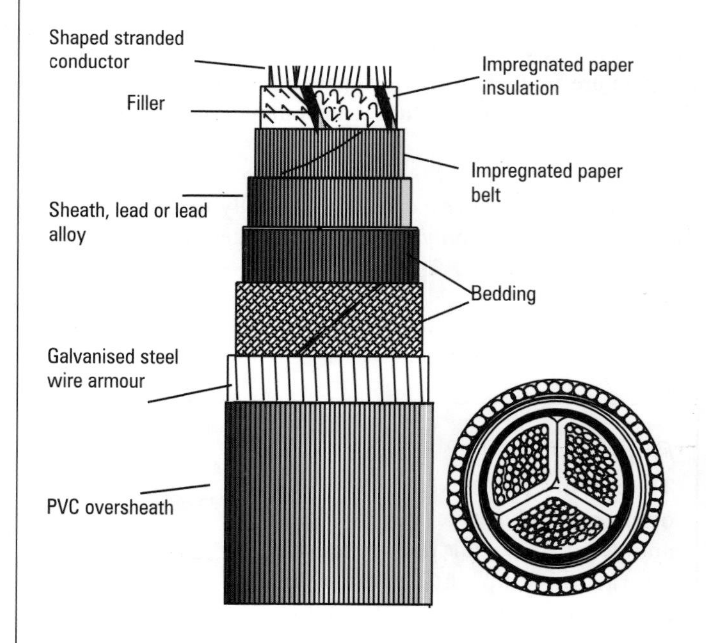

Figure 3.6 PILCSWA cable

The basis of the cable is a copper conductor with multicore cables having shaped cores. These are then wrapped with an impregnated paper tape to provide electrical insulation. The appropriate number of cores are then placed together with fibrous fillers and an overall application of impregnated paper applied to form a circular construction. The medium used for impregnation was generally a thin insulating oil but modern technology has provided a number of specialised compounds for this task. The precise formula used will depend on the intended use of the cable and the manufacturer.

An overall sheath of lead was applied but nowadays a lead alloy is applied to the circular paper bound cable. This serves to both contain the conductors and insulation and to provide protection against the ingress of moisture and damage to the insulation. If this cable is used as part of a TN-S system, it is usually this lead sheath that is used as the earthing conductor in the Public Electricity Supplier's part of the system.

Layers of paper tape are then applied over the lead and further layers of a fibrous material, such as hessian weave, are bound onto the sheath with a bituminous compound. Onto this bedding is applied the armouring for the cable protection. This may be one or two layers of either galvanised steel wire or tape.

Remember

Single core, steel armoured cables should not be used for a.c. supplies due to the inductive effect. Non-ferrous armour or unarmoured cables with suitable mechanical protection should be used. However, these may also cause problems with EMC emissions and care should be used when using such cables with particular regard to siting and screening.

The final finish to the cables is either

- bright finish bare tape or armour
- layers of hessian and bituminous compound with a whitewash finish
- PVC sheathing
- halogen-free outer sheath

dependent upon the application and the conditions in which the cable will be installed.

PIAS CONSAC

Paper Insulated Aluminium Sheathed

The basic construction for this cable is the same as for the PILCSTA cable with the exception of the conductor and sheath materials as shown in Figure 3.7. Shaped solid or stranded aluminium conductors are used and a smooth aluminium sheath is used in place of the lead sheath. In general the cable is now supplied with an overall covering of PVC and these are typically used for distribution systems. A similar cable with a corrugated seamless aluminium sheath is used for distribution up to 11 kV and a typical example is shown in Figure 3.8.

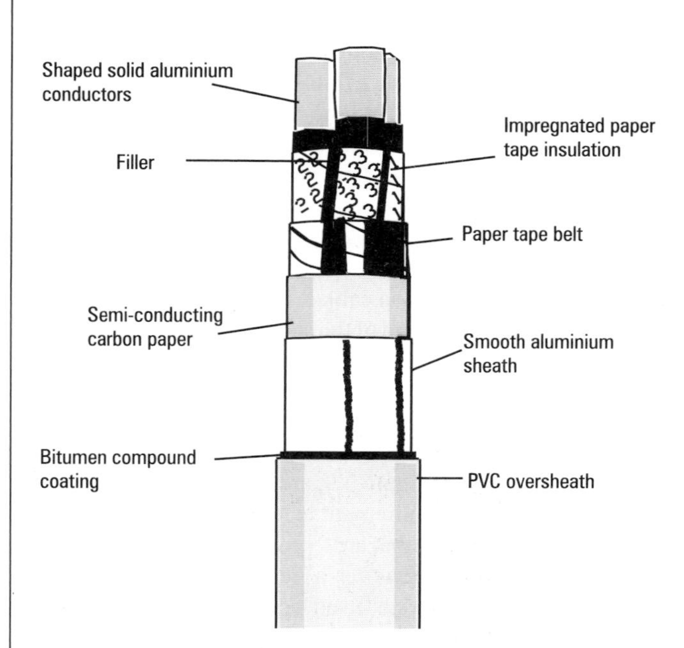

Figure 3.7 *Impregnated paper insulated, shaped solid aluminium conductors*
Smooth aluminium sheathed

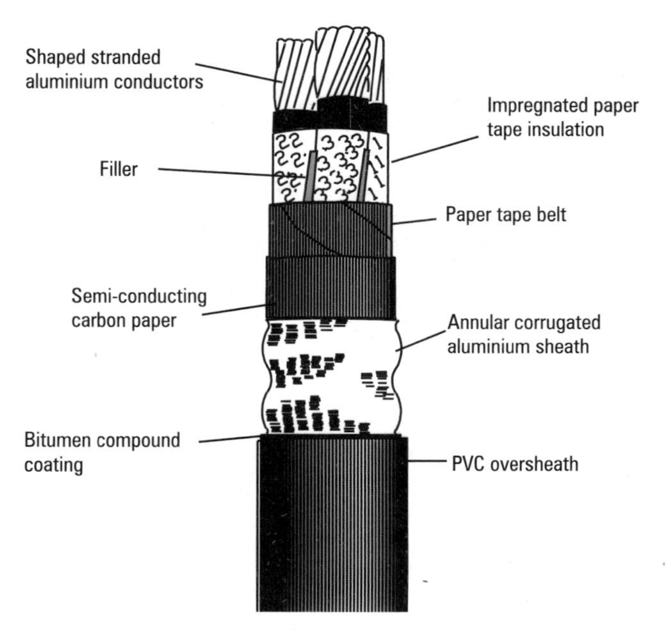

Figure 3.8 *Impregnated paper insulated, shaped stranded aluminium conductors*
Annular corrugated aluminium sheathed

HOFR and LSF

Heat and Oil resistant and Flame Retardant & Low Smoke and Fume

This is a good point at which to consider the properties of these two types of cable.

HOFR cables are usually produced with a standard cable construction but incorporating a final outer sheath that is either Chlorosulphonated Polyethylene (CSP) or Chlorinated Polyethylene (CPE). These oversheaths are classified as Heat and Oil resistant and Flame Retardant and are installed in locations where the environment is particularly harsh.

High concentrates of solvents, fuel oils and the like have a particularly adverse effect on cables. Penetration of the outer sheath can result in the spread of the solvent or fumes along the cable to switchgear and equipment, resulting in a risk of fire or explosion. HOFR sheaths provide protection against these dangers.

LSF cables, as shown in Figure 3.9, are often specified for use within buildings. This requirement is to safeguard against the health and environmental hazards produced when materials burn. These dangers occur as a result of the chemical composition of the materials used and with LSF these are chosen to reduce the dangers to a minimum.

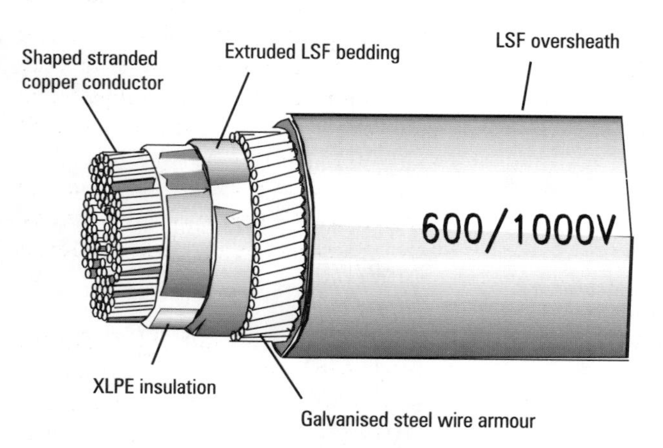

Figure 3.9 *Four core single wire armoured cable*

The materials used are normally free from halons, such as bromine, flourine and chlorine and the acid gases produced during burning are typically less than 0.5%. There are further requirements placed upon the manufacture of these cables and these are dealt with in British Standards. In particular BS 6724: 1986 deals with armoured cables for electricity supply having thermosetting insulation with low emission of smoke and corrosive gases when affected by fire.

These cables are becoming far more common and are being specified for most projects. These requirements refer to "thermosetting cables" and we know that these include XLPE and EPR cables. However, this does not mean that all of these cables are necessarily LSF and so it is important that we check that the cable we intend to use fulfils the particular specification requirements.

Installation of cables

In selecting the routes of our cables we must consider a number of important factors both with regard to the actual installation and the environment in which they will operate.

Our cables should not be installed where they would be subjected to sharp bends. The minimum bending radius will depend on the type and outside diameter of the cable, and details of these may be obtained from the appropriate British Standard. We should consider the minimum bending radius as exceptional and wherever possible we should arrange for greater radius bends. We must also remember that a cable with aluminium conductors will require a larger minimum bending radius than the equivalent cable with copper conductors.

We know from our experience with the installation of PVC cables and conduit that PVC becomes brittle at low temperatures and should not be installed unless the ambient temperature and cable temperature are above 0 °C. Should we disregard this when installing our distribution cables it may result in the PVC insulation or oversheath becoming damaged. Any PVC cables should therefore only be installed only when both ambient and cable temperatures are above 0 °C, and have been so for at least 24 hours.

Cable uses

Having considered the type of cables available to us perhaps we should now consider the uses of these cables. We shall do this by considering the cable types and typical applications for each as shown in Table 3.1

These details are only a brief guide as to some of the uses for these cables and it is not intended to be comprehensive or preclusive. We must refer to the specification and manufacturers' literature to ensure that we select the correct cable for the conditions that exist. HOFR and/or LSF cables may be required at almost any of the locations listed below. These cables should be installed in accordance with their particular merits and the requirements of the prevailing conditions and specification.

Table 3.1 Cable types and typical applications

	PVC	XPLE	EPR	Waveconal	Concentric	Paper
HV distribution over 33 kV		X	X			X
HV distribution up to 33 kV	X	X	X			X
LV distribution	X	X	X	X	X	X
Street lighting	X			X	X	
Power and substation uses						
HV distribution		X	X			X
LV distribution		X	X			X

Cable sizes

In general we find that, for high voltage systems, short circuit and fault levels will determine the minimum size of cable that we can use. A rule of thumb which covers most instances relies on the standard sizes and the cables may be selected as:

Spur feeders only:
25 mm^2

Ring main and interconnection cables:
35 mm^2, 70 mm^2, 95 mm^2, 120 mm^2

Again this is only a rule of thumb – it is not a comprehensive or preclusive guide.

Cable rating and conductor size

As we know the current carrying capacity of a cable is controlled by

- the need to dissipate the heat generated by the power ($I^2 R$) loss in the cable
- the maximum temperature at which its insulation can safely be operated
- the method of installation

Now, generally speaking, a cable in air may dissipate heat better than a buried cable. However this will depend to a certain extent on the diameter, and hence the surface area, of the cable and the ambient air temperature. Factors affecting the current carrying capacity of cables buried in the ground include the depth of burial and the thermal resistivity of the soil.

We can refer to BS 7671 and to manufacturers' published data when determining the current ratings of cables. Appendix 4 of BS 7671 lists the methods of installation and provides us with guidance on the selection of the appropriate cable size. For methods of installation not covered by BS 7671 we must rely on the manufacturers' data to make our selection.

Cable routes

When determining the route of our distribution cables there are a number of factors that we must consider. Bearing in mind the considerations that we give to the selection of cables for final circuits we can apply many of those to distribution cables. There are a few additional requirements that we should consider, for example the identification of cables used for distribution.

Except where buried or enclosed in ducting, cables should be fitted with identification discs or labels which should be attached to the cables with cable ties. Identification should be fitted within 500 mm of terminations and joints, at each draw-in pit and at least once within each separate compartment through which the cables pass. It is also good practice to fit discs or identification labels at intervals of say every 25 m along the length of the cable where possible. This practice allows the engineer to readily identify cables at all locations, which is of significant benefit during maintenance and repair operations.

When we install cables below ground we should consider the terrain and try to follow physical features of the site such as roads, fences, hedges and building lines. We should try to avoid areas of open ground or any areas which could be subject to future development. Underground cables should not be installed in areas where access for maintenance or repair may be restricted for security or any other reasons, such as land owned by others and public areas.

Where power is to be distributed around a site to feed both essential and non-essential circuits, it is desirable to keep the two circuits separate. By adopting this approach we can ensure that in the event of one cable being damaged or becoming faulty, the other may remain in service to feed the load.

Similarly where there is more than one supply to a building, these should be installed on different routes. This will also apply to high voltage ring main cables where they approach substations, or where low voltage rings approach distribution feeder pillars and so on.

Finally we must remember that we have a statutory obligation under the Electricity Supply Regulations 1988 to update the site records to show the position and depth below the surface level of any newly buried cables we have installed.

Single core power cable

When installing large, or high voltage three phase distribution systems, it is not uncommon to use two or more single core cables per phase, connected in parallel. This is often due to the physical size and weight of a multicore cable, or individual cables per phase. In such cases we should ensure that the individual cables are installed along their route in three phase groups as shown in Figure 3.10.

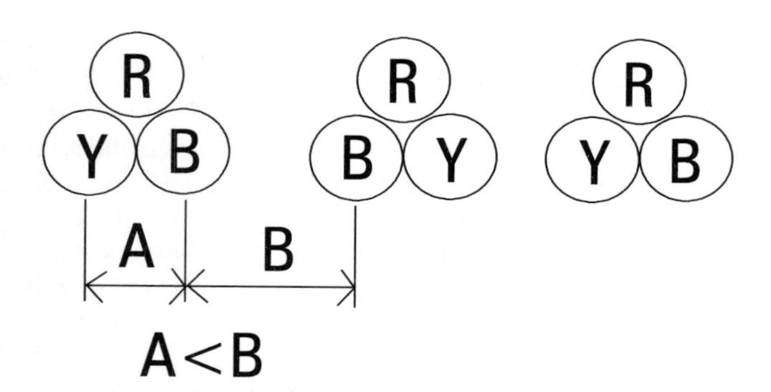

Figure 3.10 *Recommended trefoil grouping of single core cables*

Use of this arrangement will ensure that no large eddy currents circulate in the sheaths and it will provide an even distribution of current between the conductors on the same phase.

It is also important that we alternate the phase sequence within a group in order to equalise the inductance of the parallel cables in that phase, incorrect installation will not achieve our aim and is shown in Figure 3.11.

Figure 3.11 ***INCORRECT** trefoil grouping of single core cables*

Where we install single core cables in ducts, or where it is not practicable to install trefoil groups horizontally, our cables should be installed as shown in Figure 3.12.

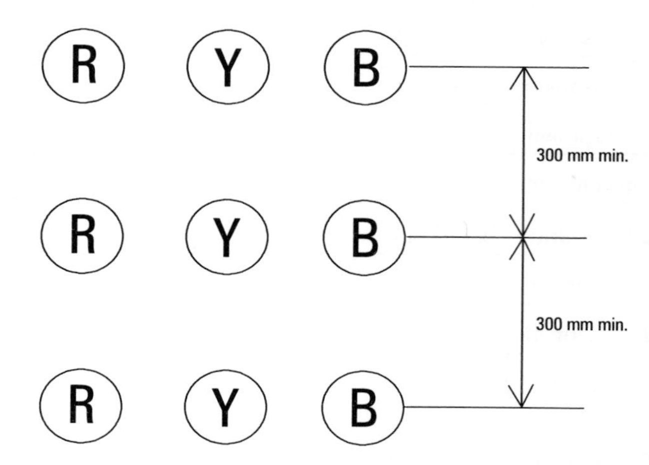

Figure 3.12 *Recommended grouping of single core cables in vertical formation*

The trefoil grouping of single core cables in vertical formation (one group above the other) is not recommended.

In cases where potentially high circulating currents in the cable sheaths and/or armour are likely and need to be suppressed, for cable rating purposes, we need to install insulating glands. The cable sheath and or armour should then be connected together and earthed at one point at the non-insulated end.

Try this

List at least two applications for each of the following types of cable:

PILCSTA
WAVECONAL
SPLIT CONCENTRIC
HOFR
LSF,SWA,XLPE

Cable ducts

Where we have to install cables in ducts beneath extensively paved, tarmac or concrete areas we will need to provide draw-pits at suitable intervals. These pits should be provided at each change of direction and at suitable intervals for straight runs. The pits should be of sufficient size to allow our cables to be correctly handled whilst being drawn in without damage, excessive bending or undue stress. It is important that some precautions are taken to prevent the pits from flooding. At the end of a duct run we should install a bell mouthed duct to ensure that we can pull the cables into the duct without causing damage to the cable.

Sealing of cable ducts

Where we enter a building from below ground level with our cables we must make sure that all entries are effectively sealed. The purpose of this seal is to prevent gas, water, oil, fire and vermin entering the building. Gas, whether occurring naturally or leaking from drains, sewers or gas supply lines, is a particular hazard. If such gases are able to penetrate the building they could create an explosive atmosphere, particularly in confined areas and around switchgear.

A number of chemical compounds are available which react to form a pliable mass around the cables. This method of sealing has the advantage of being easy to penetrate should we need to install more cables at a later date.

Special cable entry systems are available which can also provide a suitable method of sealing entries through walls and floors. The means of sealing is by passing cables through a frame into which are inserted blocks made from a special pliable material. The blocks are supplied in two halves, one for either side of the cable. Once the cables and blocks are in position, the blocks are gently compressed so that they take up any small gaps between the cable and block. This system is particularly useful for multiple cable entries.

Cable glands, joints and terminations

Joints and terminations are a fundamental part of a distribution system. They are required to perform all the functions and meet the specification requirements of the cables to which they are installed. They must have conductor connections which are suitable for the full rating of the cable and they must have sufficient insulation to equal the performance of the cable.

There are a number of methods available for the connection of lugs and ferrules to conductors, differing in their method of application and ease of installation. The two main types are the use of solder or by compression. The compression connections are easy to use and are particularly beneficial when using aluminium conductors, which require special care and attention if they are to be "soldered". The compression type of termination also removes the need for "hot termination techniques" thus the other inherent dangers of "hot working" processes are also removed. This reduces the risk of damage to conductors and insulation through over heating and the risk of fire to the building fabric and equipment inherent in such processes.

Try this

1. Briefly explain the reasons why HOFR and LSF cables are specified.

2. From manufacturers' data identify one type of LSOH and give an application for its use.

Cable joints

There are a number of methods of cable jointing available from the soldered ferrule, cambric tape, pitch filled, sweated lead, paper insulated joint to the resin cast type. The latter has become commonplace on distribution systems. These joints usually come in kit form and may be used on a wide variety of different cables, from small communication cables through to large 33 kV power distribution cables. There are two main types of joint these being "cold pour resin compound" and "heat-shrink".

The resin joints can either be of the polyurethane or the epoxy type. Epoxy resins give off much more heat than polyurethane resins when curing and therefore are only really suitable for jointing larger cables. Polyurethane resins are usually used on small cables, such as communications cables, whose insulation could easily be damaged by excessive heat during the curing process.

Modern developments in resin technology have improved their insulating characteristics and hence a smaller amount of material is now required between the conductors. This enables the overall size of the joint to become smaller without affecting its mechanical and electrical properties.

The heat-shrink joints consist of insulating tubes which are applied over a conductor connector, together with armour clamps and an overall protective sleeve. These joints have an advantage over the resin type in that they can be energised immediately the joint has been completed, they are, however only available for straight joints.

Cable termination

There are two main methods of insulating cable terminations, these are the "heat-shrink" and "cold-shrink". The heat-shrink is a well tried and tested technique but it has the disadvantage of requiring a fair amount of space around the cabling to enable the heat to be evenly applied. If we do not apply the heat evenly sleeves will not shrink evenly and not provide the required degree of insulation.

There are two variations of the cold-shrink, the float fit type which are pushed into position, and the pre-stretched type. These types are generally used for equipment with smaller terminal chambers.

There are two other principal advantages to the cold shrink method, namely, no special tools are required and no heating, filling or taping is necessary. They are suitable for use in the horizontal, vertical or angled positions.

A lot of high voltage equipment is being made which can accept a "plug-in" type of connector. The obvious advantage of this type of equipment is that it can be easily disconnected when required for cable checks or replacement of equipment. Cables may be terminated at this type of equipment by using separable insulated cable connectors.

A separable connector is an insulated device secured to the end of a cable, incorporating a female interface which engages with a male bushing on the equipment to be connected. The electrical connection between the two parts of the connector can either be by bolted joint or by sliding contact. These separable cable connectors are known as "off-load" devices and they may only be used on indoor systems with voltages between 1 kV and 36 kV.

Cable tests after installation

On completion of the high voltage cable installation we must apply a d.c. voltage of the appropriate value between the conductors, and between the conductors and the metallic sheath or armour. To establish the correct value for this voltage we need to refer to the appropriate British Standard. This test must be carried out after all our jointing has been completed, but before the cable is been connected to the system.

We must apply the voltage gradually and it must be maintained for a period of 15 minutes without breakdown of the insulation.

Other distribution systems

Having given some consideration to the cables used for distribution systems we shall now take a look at the other forms of distribution available.

The following systems are generally restricted to the distribution system within a single building. They are normally low voltage systems but high voltage versions may be installed where loading is high or particularly heavy equipment is used. In general terms this is restricted to the distribution within a single building and whilst the system may be at high voltage they are more commonly at low voltage.

It used to be common practice to install distribution systems within buildings at low voltage through steel trunking using large, single insulated or insulated and sheathed cables as in Figure 3.13.

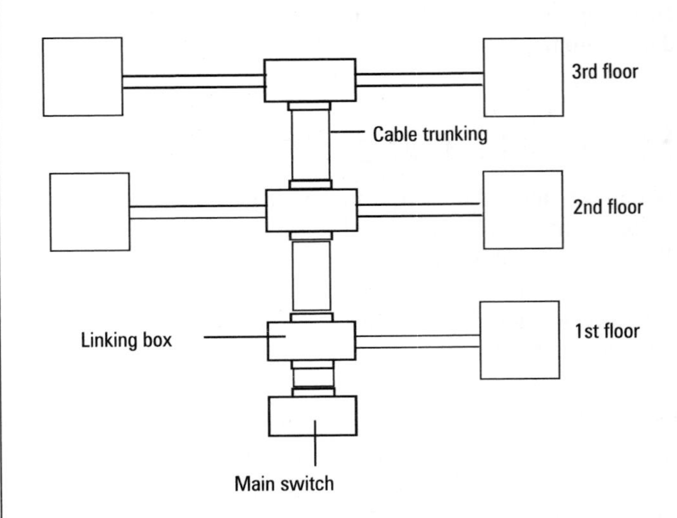

Figure 3.13 *Rising-main distribution using trunking and cable with linking boxes*

The introduction of busbar trunking allowed the installer to provide a compact distribution system and the use of take off chambers meant that equipment could be connected at any point along the route. This was particularly useful as a rising radial feeder in multi-storey buildings such as is shown in Figure 3.14.

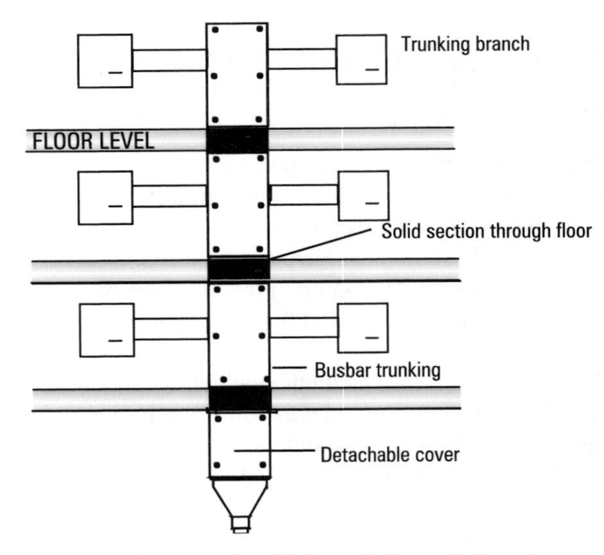

Figure 3.14 *Rising-main with laterals for each floor in a multi-storey building*

The introduction of the plug in connection to a busbar system meant that the installer could connect to the system at any outlet position without opening or de-energising the system. The inclusion of fuses at the take off point allowed overcurrent protection to be installed at the point of take-off and one such arrangement is shown in Figure 3.15.

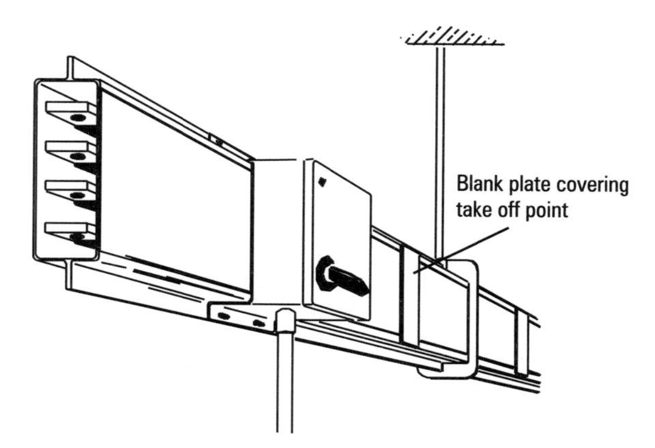

Figure 3.15 *Typical busbar distributor*

This type of system has now been refined further and rising mains and building distribution can now be installed with take off positions feeding distribution boards and the like. This type of arrangement is shown in Figure 3.16.

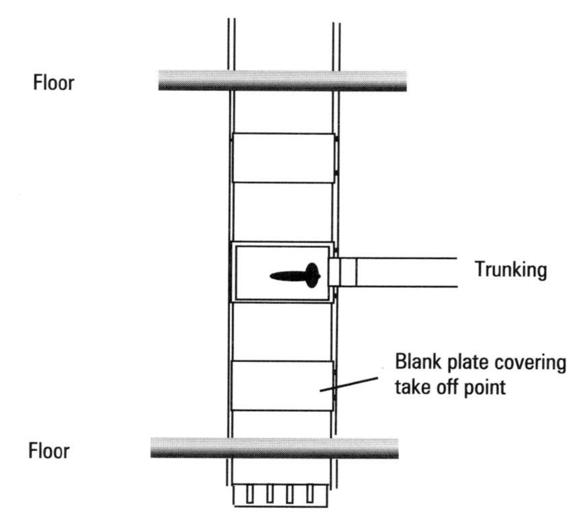

Figure 3.16 *Typical busbar riser*

When we select the type of system we are going to install we must take account of all the factors we have discussed so far and give consideration to the requirements for load balancing etc. When we are determining loading for a distribution system we are allowed to apply diversity to loading in the same way as we would to determine the total connected load to a system. Remember that we can only apply diversity on the basis of normal use. If we are aware that, for example, the lighting is going to be on continually, we cannot assume a diversity requirement.

Exercises

1. Sketch the construction of the following cables identifying each component.
   a) 4 core PILCSTA
   b) 3 core waveconal
   c) 1 core split concentric

2. Sketch a suitable distribution system layout for a 3 storey building which requires 3 No. take-off points per floor to supply tenants. The construction of the building is steel framed with solid floors, access through the building is via a single riser located centrally.

3. A distribution system comprises 6 No. single core cables laid from a substation underground and entering the building through underground ducts. It is then installed for 20 m on cable ladder to the switchroom. State the necessary precautions required for each part of the route and sketch typical arrangements at the point of entry to the building and when installed on the cable ladder.

4

Distribution Equipment

Revision questions:

1. List the factors which will affect the choice of cable for a distribution system.

2. List two advantages and two disadvantages for both copper and aluminium conductors in distribution cables.

3. What considerations must be made if single core armoured cables are to be used as part of the distribution system?

4. What tests should be carried out to ensure that a high voltage cable is suitable to be placed into service?

On completion of this chapter you should be able to:

- ◆ describe substation layouts and methods of transformer cooling
- ◆ describe methods of arc control
- ◆ state applications for current and voltage transformers
- ◆ state effects of power factor on fault levels
- ◆ state use of overcurrent and earth fault relays

In this chapter we shall consider the requirements for substations, the cooling of transformers and fault current and overcurrent protective devices associated with the distribution system. We shall also look at the use of current transformers, relays and trips and the effect that power factor has on fault levels.

We shall begin by looking at the layout of a typical substation and for this we shall consider an external substation on a consumer's property. This forms part of a ring main around the site and supplies one building in the complex.

The layout may vary dependent upon the type, size and amount of equipment contained and Figure 4.1 shows a fairly typical, single transformer substation. The items of equipment are shown as separately mounted for clarity, there are a number of units available which combine switchgear and protective devices.

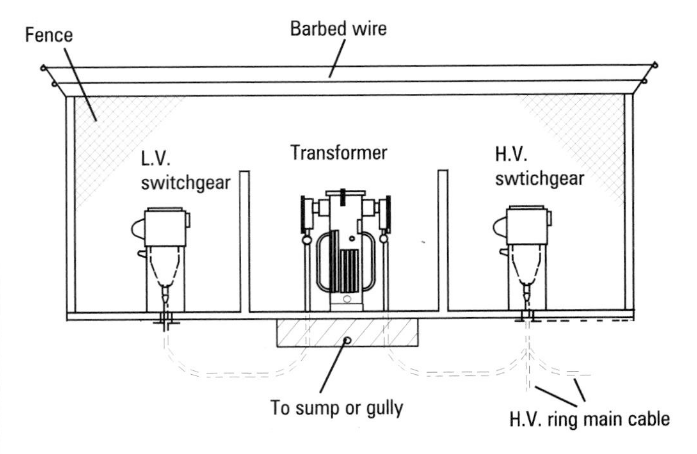

Figure 4.1 *Typical substation*

An outdoor substation does not need to be covered and so the equipment installed must be to a suitable IP rating for an exposed environment. The boundary of the substation must be clearly identified and enclosed to prevent access by unauthorised personnel. This enclosure may be a brick construction, open roofed, or a fence arrangement. Fences must be either close linked or closed and of sufficient height, fitted with deterrents to prevent people from climbing into the enclosure.

The layout of the substation shows the basic minimum requirements for an oil-filled transformer. The amount of equipment may be significantly increased depending on the complexity of the system and the extent of any monitoring required. Additional oil drainage may be required, feeding into the sump, from any oil-filled switchgear which may be installed.

We can see that the station contains the HV switchgear and the transformer. The equipment is supported on suitable bases which have access through for the cabling to each item of equipment. It is also common practice to create HV switch rooms separate from the transformer enclosure.

When the perimeter of the substation is not brick built, it should have a curb around it at least 200 mm deep. It is a common practice to then fill the base of an oil filled transformer substation with shingle, the primary function of which is to absorb and contain oil spillage.

This is sometimes done by creating a sump either within or outside the substation and a drainage system to ensure that any oil spillage is taken to the sump; this may be part of a pumped system or full transformer oil capacity. In any event the requirement is that the holding system is capable of containing or controlling the total quantity of oil contained in the transformer. Similar requirements are made of internal substations and the requirement to contain and absorb oil spillage is very important.

It is common for cable trenches to be installed and covered with concrete slabs to allow ready access to cabling without the need for major excavation, so at most it is just a question of moving gravel.

Figure 4.2 shows a typical plan of a substation and we can see that there is a good deal of space around the equipment. The Electricity At Work Regulations 1989 require clearances between equipment and between equipment and building structure to allow access to work, persons to pass and for escape. We often find that we have a situation, particularly with small switch rooms, where a second door needs to be provided to comply with these requirements, our substation is subject to similar requirements.

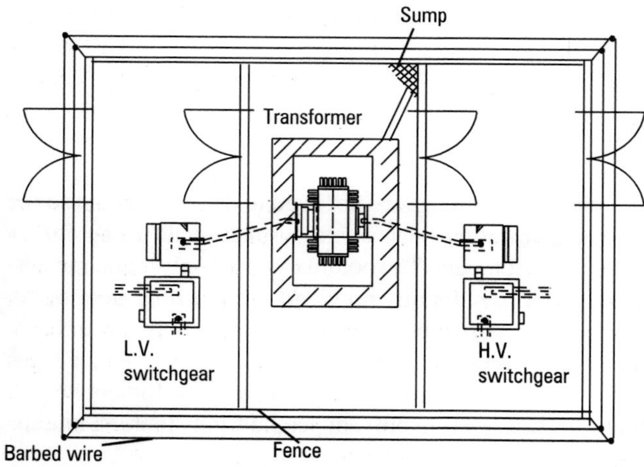

Figure 4.2 Typical plan of a substation

Transformers are large and heavy, a typical 200 kVA transformer would be in the order of 1.4 m long, 1.1m wide and 1.4 m high, weighing around 1100 kg and containing approximately 400 litres of oil. There are methods, other than the use of oil, for controlling the temperature of the transformer, these are, in the main, silicone and Dry class C types.

If we compare the types mentioned, we find that the operation losses for the oil and silicone types are about the same and the Dry class C is higher. The oil and Dry class C are about the same mass with the silicone being around 1000 kg heavier. In terms of the cost of the unit, oil is by far the cheapest in terms of capital cost, with the silicone about 40% dearer and the Dry class C 80% more expensive than the oil.

From the detail above we can see that the location of the substation should provide for easy access for delivery and replacement. The particular features of each type of transformer also tends to govern the location for each type, for example:

Oil filled

- located out of doors whenever possible
- must have a provision for oil catchment
- if used indoors may require fixed fire extinguishing equipment within the substation
- should only be used at ground level due to access, leakage and maintenance problems

Silicone

- suitable for location above ground level
- suitable for location in areas of high fire risk
- must be vented into the atmosphere usually by means of an extract and supply air provision either natural or forced

Dry class C

- suitable for location above ground level
- suitable for location in areas of high fire risk
- must be vented into the atmosphere usually by means of an extract and supply air provision either natural or forced
- must not be installed in a location where the conditions could have a detrimental effect on the windings, for example outdoors or damp and humid locations, as the windings are exposed to the atmosphere

It can be seen that there are a number of factors which we must consider when selecting the location of our substation and the transformer we shall use.

It is a common practice to locate the HV and LV switchgear outside of the transformer location in a separate room, or rooms, as shown in Figure 4.3. Should a substation contain more than one transformer then "blast walls" should be installed between them.

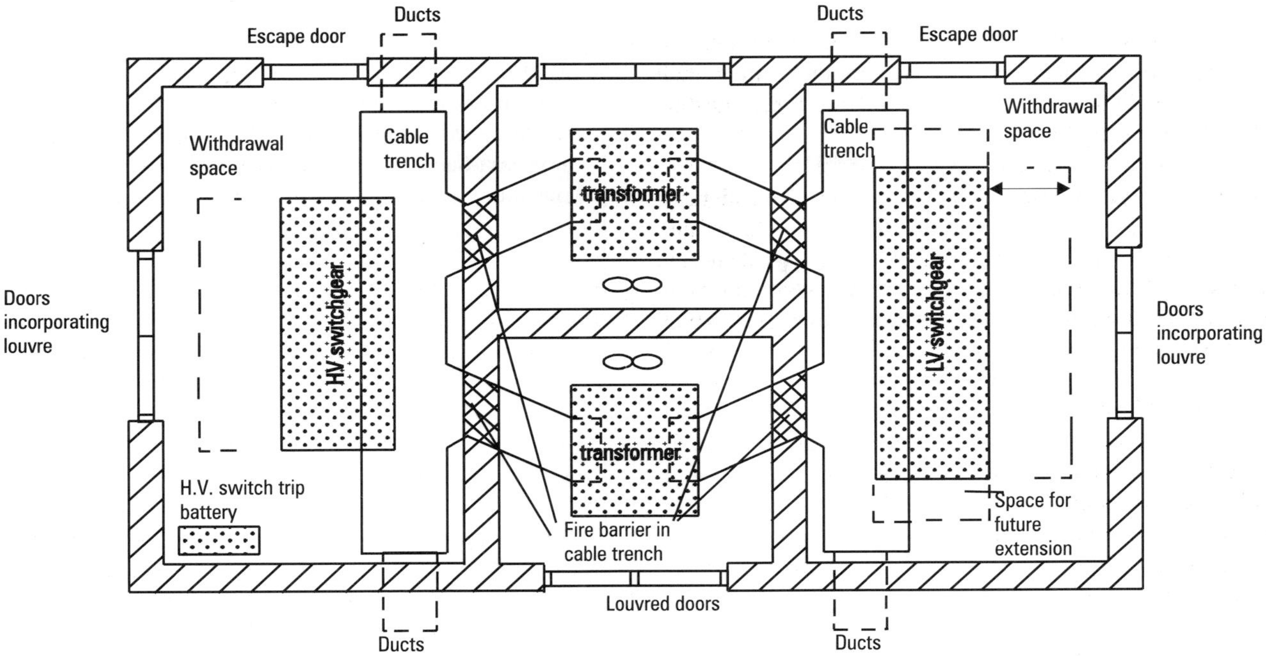

Figure 4.3

Try this

A multi-storey business development is to have all the major plant located in roof plant rooms. In view of this and for reasons of building and ground space it is decided to have a roof-mounted substation of 1000 kVA. State which type of transformer you consider most appropriate in this location and give the reasons for your choice.

It is normal for these walls to extend at least as high as the top of the cable box on the transformer.

We are now going to look at the types of switchgear and control equipment we use for our distribution system. These fulfil the same function as the control equipment used elsewhere in the installation. The main difference is the level of energy they have to handle when carrying out their function. Because of this the equipment is usually much larger, more robust and more complex. For example, arc control becomes more significant at HV with levels of up to 1000 kVA being quite commonplace and fault levels, as we saw in Chapter 2, at around 250 MVA.

Whenever we open a circuit "on load" the result of the interruption of current flow produces an arc. This is basically due to the reduction in pressure holding the contacts together immediately prior to the contacts opening, and this results in an increase in resistance. This causes a rise in temperature due to the I^2R losses and this in turn gives rise to thermionic emission from the contact surfaces, rather like splashes from boiling water in a saucepan, which tends to help continue the flow of current. As the contacts separate, a voltage appears across the gap which now has a higher resistance causing further ionisation.

These effects will produce sufficient electrons to continue the current flow across the gap and thus sustain an arc. Providing the potential across the gap and the temperature are high enough, as is usually the case in HV systems, the surrounding medium will contribute electrons to the arc due to ionisation and particle collision.

If we are to extinguish the arc each of these factors must be overcome. Now let's consider the situation where a fuse is used to open circuit in the event of a fault. Not only will the current be high but the effect of melting and vaporising the fuse element fills the area around the "separation" with molten metal. As a result the method of containing this effect must be suitable for the energy to be dissipated.

The forming of an arc results in the release of a considerable amount of energy and as we know the principle of any explosive device is the sudden release of high levels of energy. We can see therefore that the arcing process also produces an explosive force and considerable pressure is developed which needs to be contained.

This situation is further compounded by installations with low power factors. We can regard the power factor of a system, or circuit, as a measure of the energy stored within it. A system with a power factor of unity will contain none, whilst a system with a power factor of 0.6 will contain a considerable amount of stored energy.

We know that an inductive load when disconnected from the supply tries to maintain current flow by the release of the stored energy (back emf) and the arcing at switches controlling fluorescent lamps illustrates this well. The discharge of this stored energy is very rapid and one of the effects of this is that the first peak of the current cycle, following disconnection of the supply, may be considerably higher than the normal peak current value. This, of course, will assist in creating and maintaining the arc across our opening contacts, which in turn will affect the ease with which the arc can be extinguished.

Table 4.1 shows the typical values of power factor and the ease with which short circuit faults are dealt with.

Table 4.1

Power factor	Degree of difficulty	Typical location of fault
1.0	Easy	
0.9	Not so easy	Most circuits
0.8	Becoming difficult	Near transformers
0.7	Difficult	Near transformers
< 0.6	Very difficult	On or near transformer terminals

The operation and construction of various types of protective devices have been discussed in previous publications in this series. We shall look at the types of protective, isolation and control devices used in relation to the distribution system beginning with the circuit breaker. It is worth noting that the methods of arc control discussed for circuit breakers are also employed in the switches used to control the distribution system.

Try this

State briefly the cause and effect of an arc produced when opening a circuit under load and list the factors which contribute to the arc.

As we have seen there is a fair amount of energy to be dissipated and a considerable force produced when an HV arc occurs. The principle of operation for the HV circuit breaker does not vary much from that of the LV device. The additional requirements for arc control do give rise to some changes in construction and operating technique. We shall consider some of the options and look at the basic method of operation and arc control.

Oil filled

In this instance the principal of arc control incorporates immersion of the contacts in oil. The contacts are usually contained in an explosion pot, as shown in Figure 4.4, constructed to aid the extinguishing of the arc. When the contacts are opened an arc is drawn through the oil, an insulator, and gases are produced, with a little oil being carbonized in this process.

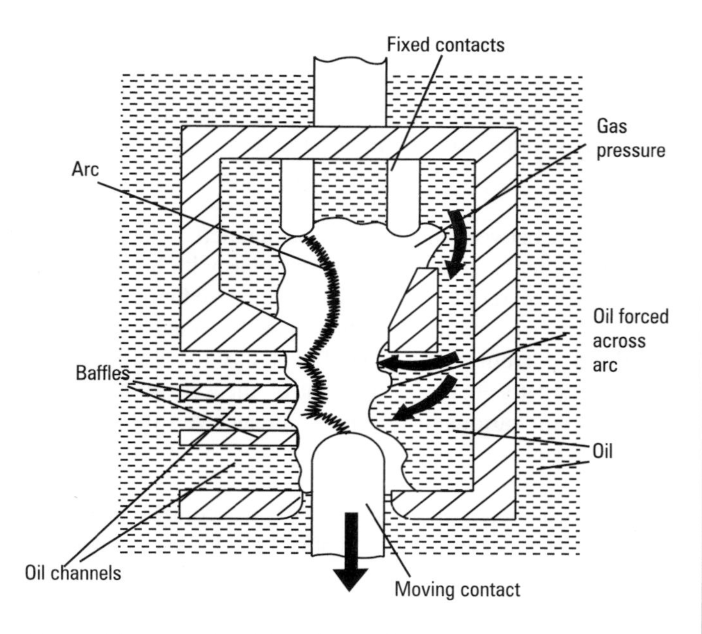

Figure 4.4 *Circuit breaker explosion pot action*

The pressure caused by the gas is used to force the oil around within the explosion pot, and across the arc, helping to extinguish it, the oil prevents the arc and its by-products from entering the atmosphere. In some breakers the oil is pumped across the contacts to extinguish the arc rather than rely on the pressure of the gas produced.

Remember

The oil used in transformers and circuit breakers, although contained in flameproof enclosures, is a fire and health hazard.

Extreme care should be taken with the handling of oil during maintenance.

Air blast

Figure 4.5 shows the much simplified layout of an air-blast circuit breaker. As the contacts open a blast of compressed air, at around 20 bar, is forced across the arc rapidly extending its length and this quickly extinguishes it. The precise design of these breakers does vary, dependent upon the manufacturer and operational requirements. Some, for example, use a spring mechanism to open the contacts whilst others use the compressed air to open the contacts and extinguish the arc.

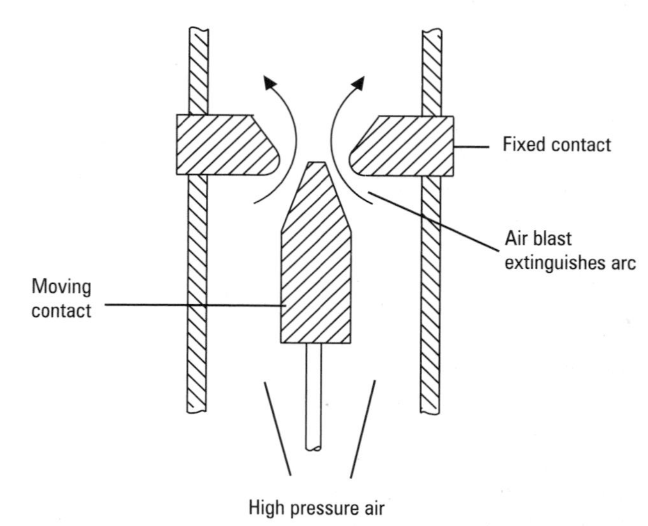

Figure 4.5 *Air-blast circuit breaker*

Obviously there is considerably more additional equipment required to operate this type of breaker. Air reservoirs, compressors and operating control gear and the like makes this type more complex and therefore attracting a higher initial cost. On the plus side however the arcing time is shorter, maintenance is cleaner and easier and the fire risk is negligible.

Gas

The operation of this circuit breaker is very similar to that of the air-blast circuit breaker except that a gas, such as Sulphur Hexaflouride, is used. This is a better insulator than air or oil, is non toxic and non-flammable, inert and stable. The gas, at around 4 bar, surrounds the contacts, as they open a blast of gas at around 16 bar is forced across the arc.

Certain precautions are necessary with this type of breaker such as the installation of heaters in certain locations to prevent the gas from liquefying at low temperatures, say 9 or 10 °C. As the gas is expensive, it is usual to pump it into a storage tank during maintenance and for checks to be made regularly for any leakage.

Vacuum

The vacuum circuit breaker is relatively maintenance free as the contacts are contained within a sealed vacuum container with the moving contact connected to the outside through a bellows arrangement as shown in Figure 4.6. The contact surfaces are in the form of flat discs and being contained in a vacuum there is no other medium to cause ionisation.

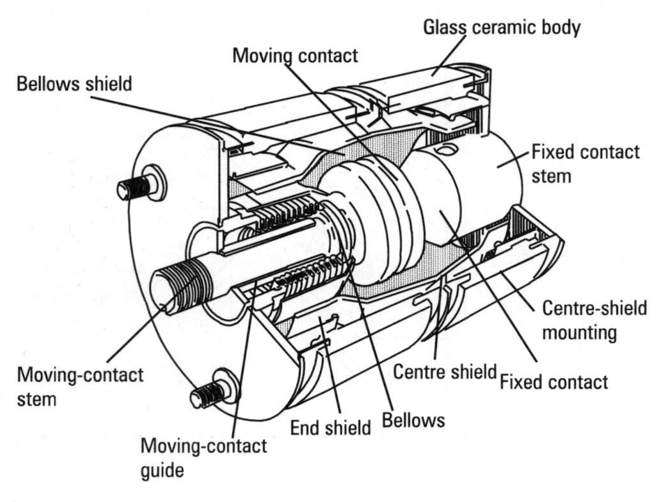

Figure 4.6

The result is that the arc is extinguished the first time the current passes through zero on the waveform, with minimum damage to the contact faces.

Moulded case circuit breakers

We are familiar with the construction of the MCB and the MCCB is really a more sophisticated version. The ability to break larger fault currents with this device is as a result of a more refined contact and arc control system. We can see from Figure 4.7 that one method of achieving this involves an additional set of contacts, the arcing contacts, with arc runners and an arc chute with splitter plates.

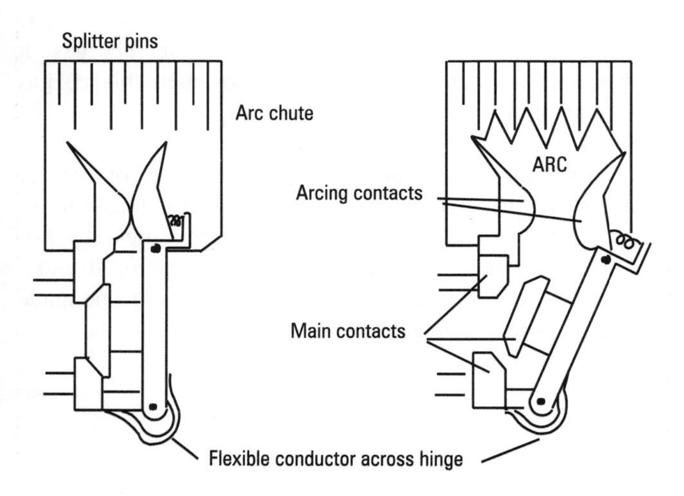

Figure 4.7

When the main contacts separate, the arcing contacts remain in together and the arc is only initiated when the "arcing pair" separate. The vaporisation and heat distortion to the contacts are confined to the arcing pair. As these do not need to carry the load current during normal operation they can be made of a material such as carbon. The arc is drawn out along the route of the arc chute and the splitter plates extend the arc to create a longer run within a more confined physical space.

Alternative manufacturers' designs deal with arc control in a number of different ways but the common feature is the design of the arc chute. It is the function of the chute to increase the length of the arc as rapidly as possible over the greatest possible distance. The more able we are to do this and control the energy released the higher the fault current we can disconnect.

The balance between the thermal and magnetic operators does not really affect the fault current control as with high currents the magnetic part of the device should operate far quicker than the thermal trip, which is better placed to deal with lower overload currents. Most MCCB's have an adjustment incorporated to enable us to control the sensitivity of the device to suit various loads hence a greater versatility from a single device.

Having considered the methods of arc control in circuit breakers we should now remind ourselves of the way in which the various fuses achieve their "arc" control.

Semi enclosed (BS 3036)

Figure 4.8 shows a typical BS 3036 fuse carrier. The method of arc control consists of creating an air gap between the two ends of the molten fuse. The operation being such that the fuse element carries a high current and as a result becomes so hot it melts, in fact much of the material vaporises.

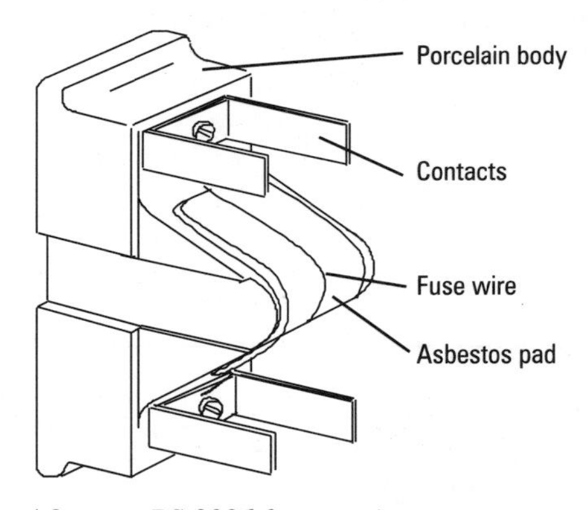

Figure 4.8 *BS 3036 fuse carrier*

The products of this reaction are contained by the asbestos pads, one on the fuse carrier the other on the base, absorbing the heat and molten material and allowing the separation of the two ends to be sufficiently far apart so as to extinguish the "arc". In certain types of fuse carriers the fuse element is surrounded by ceramic material with the "port hole" closed by an asbestos pad in the base.

Cartridge type (BS 1361)

These employ the same fuse wire principle as the rewireable fuse only the fuse element is contained within a ceramic body as shown in Figure 4.9. With this type when the fuse element vaporises the scattering metal particles are contained within the ceramic body. Again distance is the factor that extinguishes the arc but in this case the device offers far less risk of fire or explosion.

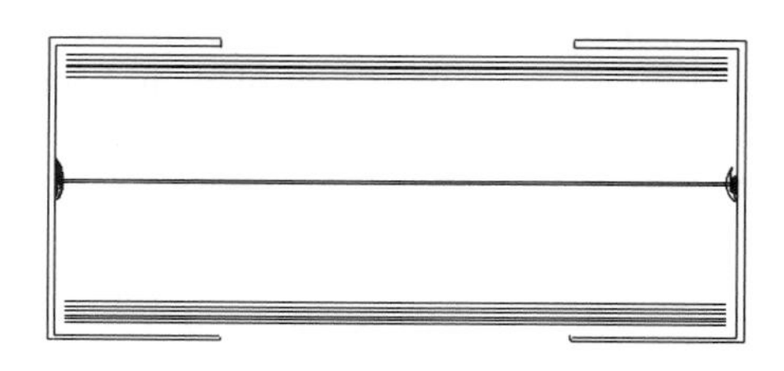

Figure 4.9 BS 1361 cartridge type fuse

High breaking capacity (BSEN 60269-1)

A typical construction of this type of fuse is shown in Figure 4.10. The use of multi elements ensures that once the first element fails the remainder follow rapidly, in an avalanche effect, as a result of the increased load placed upon them. The silica sand flows into the gap created by the disintegrating elements and extinguishes the arc. The sand is usually melted to become almost a glass with the high fault currents that these devices can disconnect.

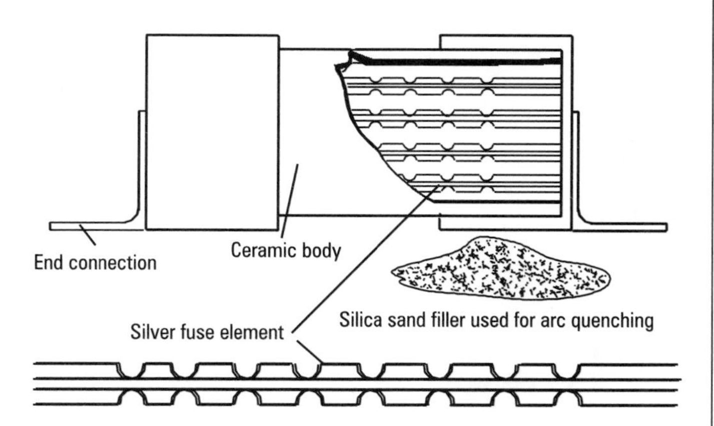

Figure 4.10 HBC (High Breaking Capacity) fuse

At this point it is worth mentioning that in the case of HV control equipment it is not uncommon to find oil filled fused switches used where the fuses are immersed in oil. Other types of fuses are used such as the Trip Pin fuse where an explosive charge is detonated on operation of the fuse which causes a pin to penetrate the end cap. This action is used to trip a three-phase switch so disconnecting all phases in the event of a fault. Other types use the pin principle as an indicator to show that the fuse has operated and a typical section is shown in Figure 4.11. These types are special in their use and operation and often feature in the equipment used by the supply companies.

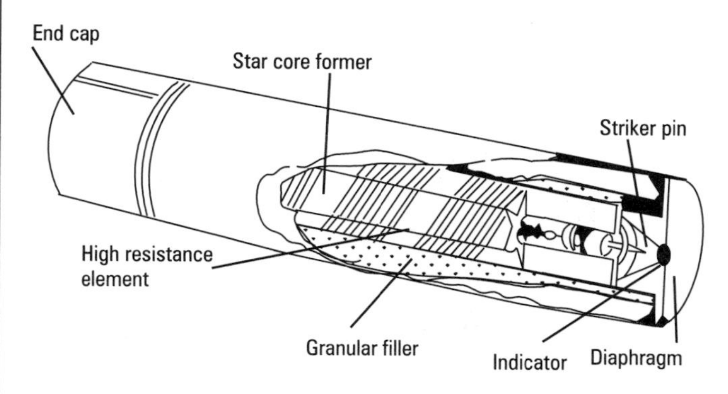

Figure 4.11 Pin indicator type

Try this

Describe the method of arc control for two different types of circuit breakers and two types of fuses giving advantages and disadvantages for each type.

The devices we have considered so far tend to protect our distribution system from short circuit currents and it is now time to give some thought to the protection against earth fault currents.

As a result of the high voltages and currents that we encounter in our distribution system it is not practical in many instances to provide direct monitoring of currents and voltages. It is a common practice to use current and voltage transformers to allow us to measure and monitor the system. These devices are also used in the tripping circuits for the operation of our switchgear.

The principle and operation of these transformers have been covered in other titles in this series and so we shall not be dealing with this again in this chapter. We shall consider the use of these transformers and their function within the system. Figure 4.12 shows a simplified circuit breaker with CT and VT installed.

We are familiar with the operation of a motor starter and the use of coils and open and closed contacts for control. The operating principle of the devices used to control our distribution equipment is very similar, with various devices used for sensing and tripping. The thermal and electromagnetic devices used for control of our LV equipment are not always suitable in an HV application where precise discrimination and operation are required.

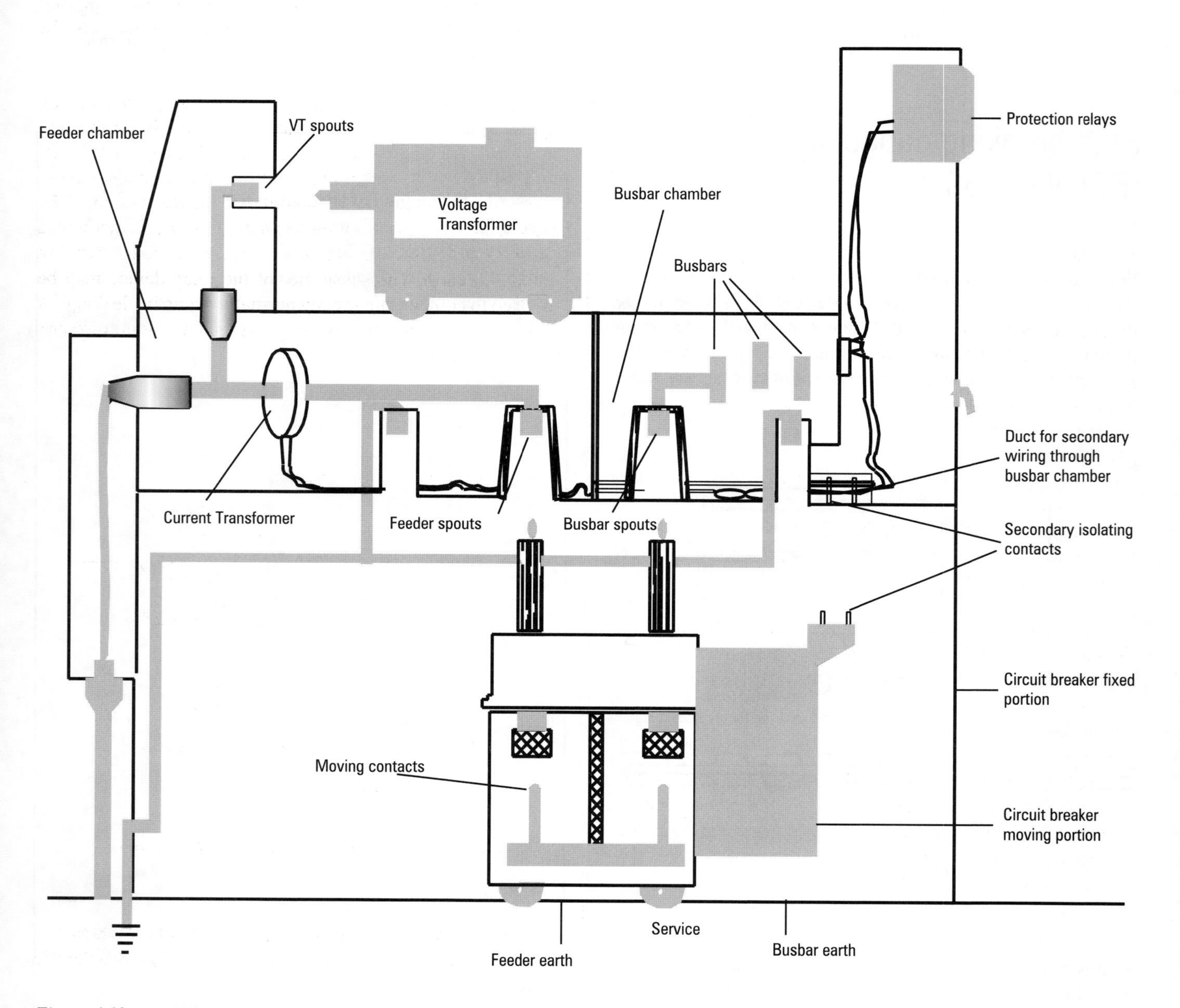

Figure 4.12 *H.V. circuit breaker*

Let us consider a typical device used for this purpose, i.e. the induction relay. The basic circuit diagram for this is shown in Figure 4.13 and we can see that the supply to the relay is derived from a current transformer in the circuit to be protected. The way in which the relay operates is similar to that of an induction wattmeter and the interaction between the magnets cause the disc to rotate.

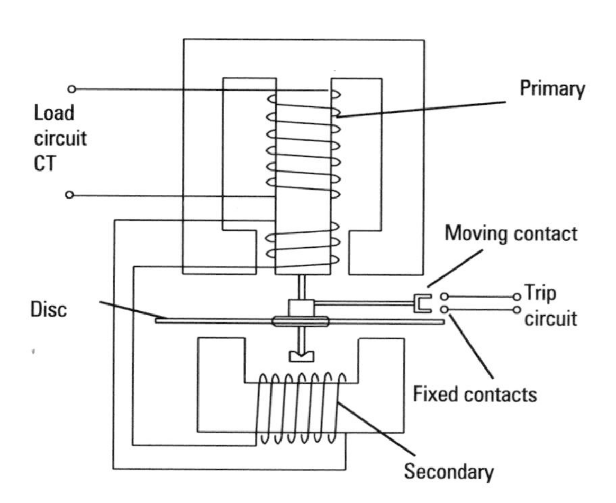

Figure 4.13 *Induction type overload relay*

Spiral springs are used to hold the disc stationary at normal operating current, of about 5 amperes, in the primary circuit. Excess current drawn in the main circuit is relayed to the primary circuit and causes the disc to rotate. After a set rotation, usually 180°, the moving contact connects the two fixed contacts together and operates the tripping circuit. The more severe the fault the faster the disc will rotate and so disconnect the circuit.

It is usual for the primary coil to have tappings, so that a variety of trip currents are available for selection. To complete the versatility of the relay the time period may be adjusted by varying the start position of the contacts thus increasing or decreasing the trip time.

If a current transformer (CT) is incorporated in the earth path then we can sense earth fault currents in the same way as we detect overcurrents. The sensing coil will need to have a higher sensitivity as little current is expected to flow during normal operation.

A simple layout for such a system is shown in Figure 4.14. This relay sensing coil will detect a fault to earth at any point in the system since any fault current will flow through the earth current transformer. The use of this method will provide us with unrestricted earth fault protection.

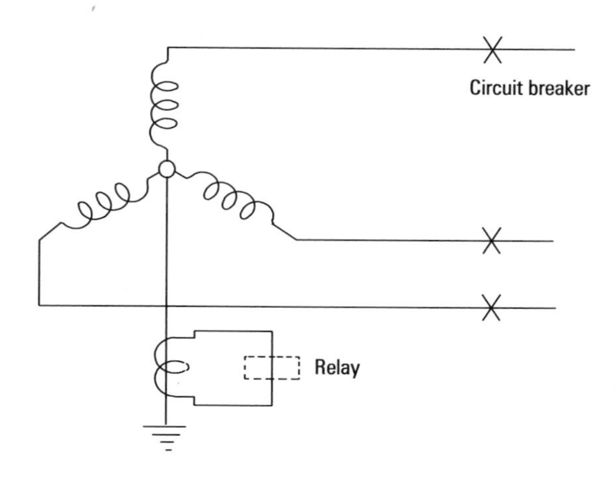

Figure 4.14 *Simple sensing coil schematic*

Where the system comprises mostly delta connected transformers the detection of earth fault currents must also be considered.

We can install a sensing device, of which a simplified version is shown in Figure 4.15, which will provide earth fault protection for the delta connected system. We can see that this device will operate on an earth fault on the system but will not be operated by any phase imbalance in the system. The transformer core will have no circulating magnetic flux with a phase imbalance. A flux will however be produced if there is a leakage to earth. The sensitivity of the relay device may be adjusted to compensate for any normally occurring leakage. A time delay mechanism may be incorporated to minimise any nuisance tripping.

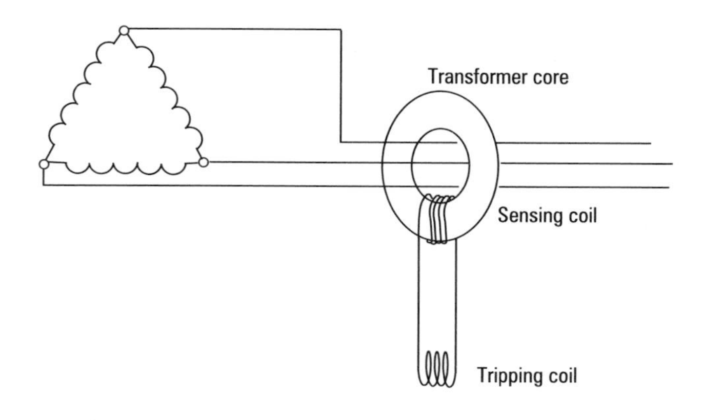

Figure 4.15 *Earth fault protection of a delta system*

This means that the protection provided will be restricted to phase and earth faults. These are often referred to as Balanced Current Relays.

Exercises

1. Sketch and label a typical substation layout with two transformers.

2. State the effect of a low power factor on fault levels in distribution systems.

3. Sketch and label the contact compartment of a vacuum circuit breaker.

4. State the application for voltage and current transformers in monitoring and protecting against earth leakage in distribution systems.

5. Sketch and label a simple induction type overload relay.

5

Earthing

Revision questions:

1. Which Statutory Regulation deals with the requirement for working space around substation equipment?

2. List the three principal methods for controlling high voltage transformer temperature.

3. List the four main methods of arc control used for high voltage switchgear.

4. List the principal devices used for protection against overcurrent and briefly describe the method of arc control used in each.

5. Briefly describe the one method used to provide protection against earth fault currents in high voltage systems.

On completion of this chapter you should be able to:

- state the hazards arising from earth leakage
- describe and explain earth fault loop impedance
- describe testing of earth electrodes
- describe methods of providing earth leakage protection
- explain voltage gradients in relation to earth electrodes
- state the effect of fortuitous earth paths

In this chapter we shall consider the requirements for earthing. This applies throughout the system and we shall pay particular attention to the earthing of consumers' installations. We shall also consider the requirements for earth leakage protection and how this can be achieved.

Statutory and non statutory regulations require the protection of persons, property and livestock from uncontrolled electric current flowing in an electrical installation. As this includes electrical discharge to earth the reasons for earthing the system becomes more apparent. If we can provide a reliable connection between the system and earth with a low impedance then the tendency will be for fault currents to take this path. We can then control the leakage current by providing a means of automatic detection and disconnection thus reducing the risk.

Try this

Before we carry on with this chapter familiarise yourself with the requirements of The Electricity Supply Regulations 1988, Part II, Connection with Earth.

The Electricity Supply Regulations 1988 requires the supplier to ensure that the system is connected to earth and this may be done in a number of ways. The most common is an earthing conductor connected between an earth electrode and the star point of the supplier's step down transformer. This is also the point of connection for the neutral conductor of the system and so achieves compliance with the requirements of The Electricity Supply Regulations 1988.

When the high voltage sections of the transmission and distribution are connected to earth it is normally via an arc suppression coil. Overhead line transmission and distribution cables are liable to direct lightning strikes and transient over voltages through inductive coupling. The energy produced as a result needs to be dissipated. Underground cables may also be affected by lightning strikes but this is generally as a result of resistive coupling through the mass of earth.

For our part we shall consider the connection of the system to earth. We shall begin at the first point of earthing on the low voltage part of the system, the star point of the transformer. The three most common systems available are TT, TN-S and TN-C-S. Each of these systems indicates a particular method of providing the required earthing.

Each of these system types will have a particular Earth Fault Path, the purpose of which is to allow earth fault currents to return to the source with the minimum of danger. We shall look briefly at each of the three systems and consider the way in which we achieve the safe return of earth fault currents.

TT system

In the TT system the supplier ensures that the star point and neutral of the system transformer are connected to earth. The earth return path to this transformer earthing point is via the general mass of earth. An earth electrode is installed as part of the consumer's installation. The intention is then to use the mass of earth as the return conductor as shown in Figure 5.1.

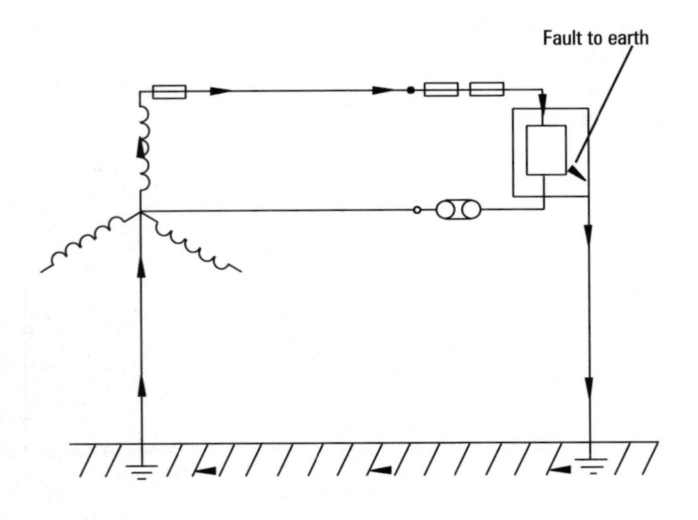

Figure 5.1 Earth Fault Loop for TT system

Now the mass of earth may have a high impedance, let's suppose that this is 2300 Ω. If the supply is at 230 V and, if we ignore the impedance of the other conductors, we can calculate the current flow around the loop to be $230\ V \div 2300\ \Omega = 0.1\ A$. A current of 100 mA is sufficient to be fatal so some additional means of protection against electric shock must be incorporated. We shall consider the means of providing this protection later in this chapter.

TN-S system

With this system we have a return path from the consumer's installation to the source provided by the supplier. This is a separate conductor from the live conductors of the system and is usually the sheath or armouring of the cable. In the event of an earth fault the current flow in the system would be as indicated in Figure 5.2 and the impedance of the conductor is normally low enough to ensure that a high current flows in the loop.

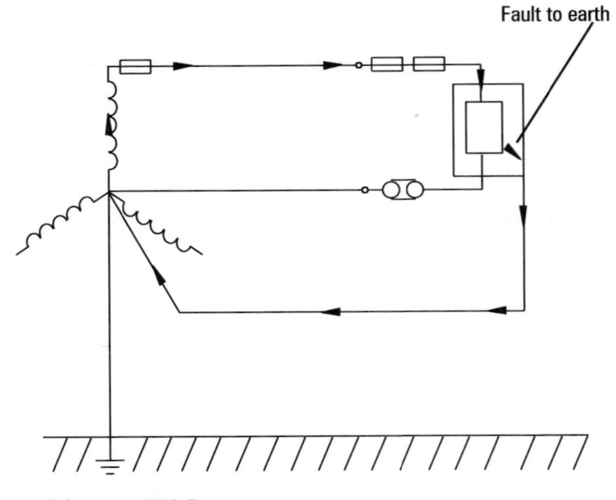

Figure 5.2 TN-S system

TN-C-S system

In this system, as with the TN-S system we have a return path from the consumer's installation to the source, provided by the supplier. In this case that is via the supply neutral conductor to the consumer's intake position at which point the supplier provides an earth terminal. From this point through the consumer's installation it is normal for the earth connections to be made by separate conductors. The current flow around the loop in this system is shown in Figure 5.3.

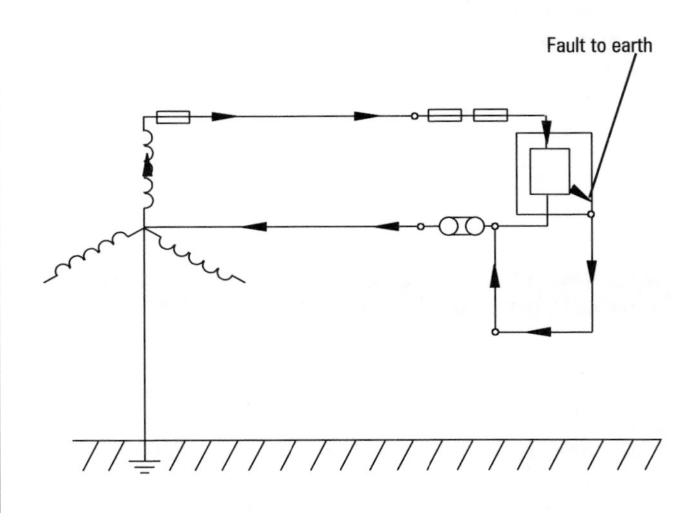

Figure 5.3 TN-C-S system

Those are the three principal systems generally available from the suppliers. In order to provide protection against the risk of shock, fire and burns we need to ensure that the supply is automatically disconnected in the event of a fault occurring on the system. There are exceptions to this requirement detailed in BS 7671 and one of these is related to the risk involved in removing the supply from certain items of plant or equipment which would result in greater danger.

An overhead electromagnetic crane in a steel works is one such example. In the event of an earth fault occurring on the system, which causes the supply to the crane to be isolated, then several tonnes of steel could fall without warning. In such circumstances we would not want automatic disconnection because of the other risks involved.

We would be expected to provide a system of earth fault monitoring which operates audible and visual warnings in the event of a fault occurring. This will allow the operators to safely position and park the crane before turning off the supply. The reason for the fault would then need to be established and rectified. It is common practice to supply such equipment with its own dedicated circuit separate from any other protective device.

Earth fault loop impedance

The earth fault loop path is made up of the phase conductor from the supply transformer, the supply cables and the consumers installation to the point of the fault. If we assume a fault of zero impedance, the rest of the loop is made up of the circuit protective conductor, the earthing conductor, the earth return path (dependent on the type of system) and the transformer winding as shown in Figure 5.4.

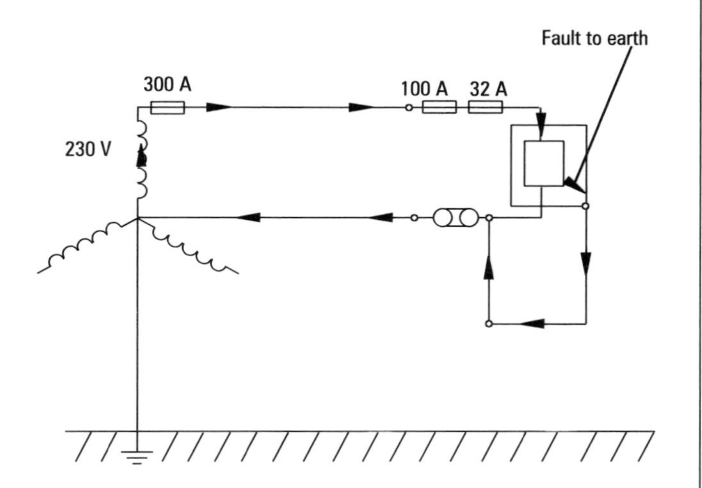

Figure 5.4 *TN-C-S system*

Providing the earth fault path is made up of conductors of low impedance then the total impedance of the loop will be low. For example if the total impedance of the loop is 0.9 Ω and the voltage to earth is 230 V then the current that will flow in the event of a fault of 0 Ω between phase and earth will be

$$230\text{ V} \div 0.9\ \Omega \quad = \quad 225.56\text{ Amperes}$$

The lower the earth fault loop impedance the higher the current and the higher the current the quicker the disconnection time will be. We not only have to disconnect the circuit from the supply but do so quickly, in as little as 0.2 seconds in some locations. The higher the current flow the faster the device will disconnect and the lower the earth fault loop impedance the higher the current.

We shall not deal with the process of determining the suitability of the earth fault loop impedance for protection by the use of the protective device, fuse or circuit breaker in this book. This subject is covered two other books in this series, Stage 1 Design and Stage 2 Design.

If we consider the above example and connect this circuit to a TT system, as shown in Figure 5.5 then we have an altogether different situation.

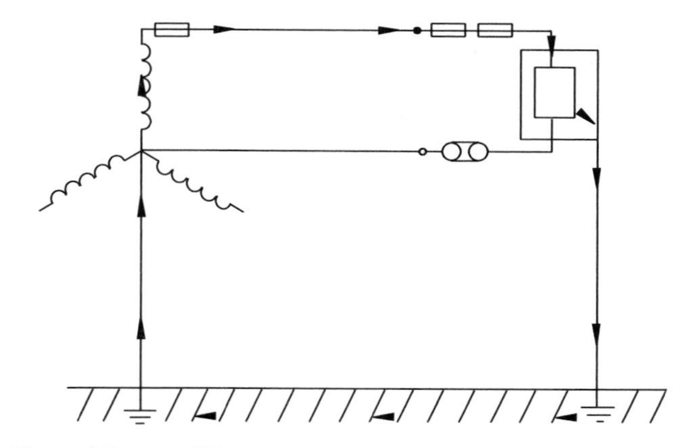

Figure 5.5 *TT system*

By reference to BS 7671 we can determine the disconnection time for say, a single phase circuit in an agricultural installation protected by a 32 A BS EN 60269-1: 1994 (BS 88) fuse, in these conditions the maximum disconnection time will be 0.2 seconds. We can see from the maximum values given in BS 7671, Section 605 that if the total earth fault loop impedance is above 0.92 Ω then this fuse will not disconnect within 0.2 seconds.

In practice we will be very lucky if the earth electrode resistance is less than 2 Ω and it is likely to be well above 10 Ω. In which case the protective device will not operate in time to provide protection against the risk of shock from indirect contact.

In these circumstances we need to fit a Residual Current Device to provide shock protection. In our above example we can determine whether the RCD will be suitable by the use of the formula

$$R_a \times I_{\Delta n} \leq 25\text{ V}$$

where R_a is the earth electrode impedance and $I_{\Delta n}$ is the rated tripping current of the RCD.

By a rearrangement of this formula we can determine the maximum rating of RCD that we can install for any measured electrode resistance, as 25 V $\div R_a = I_{\Delta n}$

In our example if the electrode resistance was found to be 50 Ω then we can use an RCD rated at no more than 500 mA.

As the RCD is fitted to the front end of the installation, providing protection for all the circuits, it is not desirable to install a typical 30 mA device as this can result in considerable nuisance tripping. It is better to install a higher rated device, with a time delay if necessary, and provide a second device where closer protection may be required. The levels of tripping current alone will not afford discrimination between the RCDs in such circumstances. Without the time delay it is likely that, with values in excess of 50% of the upstream device, it will be this device which operates first. The important factor is the time delay feature of the upstream device as this will allow the lower rated RCD to operate first.

Remember

With the TN systems (TN-C-S and TN-S) the supplier provides a Main Earthing Terminal (MET) for the consumer's use. With a TT system there is no facility provided by the supplier for earthing and the consumer is responsible for the earthing of the installation. We, as the electrical contractor, have a statutory obligation to ensure that the consumer's installation is suitably earthed.

Try this

1. Draw a circuit diagram of the earth fault loop path for TT, TN-S and TN-C-S systems identifying each component.

2. You carry out an earth fault loop impedance test on an existing TN-S installation to establish Z_e. The reading obtained shows Z_e to be 7 Ω. State what action you would take
   (a) to protect the user of the installation immediately
   (b) to effect a long term remedy
   giving reasons for your actions in each case.

Installation and testing of earth electrodes

We have looked at a number of factors related to TT systems and the earthing of the consumer's installation by use of an earth electrode. This is a good time to consider the construction, installation and testing of earth electrodes. The same criteria apply to the supplier's earthing electrodes and those for lightning protection of structures. The main difference is the level of fault current to be carried by lightning protection electrodes and, as a result, the electrode resistance must be low.

In some areas, particularly the rural ones, the TT system is quite common. In city areas TT systems are less common as most services are underground cables on a TN system. However, the suppliers are becoming more and more reluctant to allow the TN-C-S earth to be used outside the equipotential zone of the installation. This means that any detached building, such as a garage or workshop, must be separated from the main earthing terminal and provided with an electrode, effectively making the outbuilding its own TT installation.

Earth electrodes

The sole function of the earth electrode is to provide a reliable connection with the mass of earth. To do this there are a number of types of earth electrode available and a number of factors which affect which type and size of electrode we need to install. Remember that we need a connection to earth that will offer sufficiently low resistance to operate our protection device.

The main factor that we need to consider is soil resistivity and this will vary depending on the make up of the soil. It is subject to the moisture and chemical content of the soil and will be prone, at shallow depths, to seasonal variations, the long wet winter or hot summer. One method of overcoming these problems is to ensure that the electrode is sufficiently deep to make contact at a level where these seasonal changes do not occur.

The most common electrode is the earth rod which is driven into the ground without the need for large excavation works and the like. There are two main methods used to provide improved electrode contact, the deep driven rod and the use of parallel rods. Details on the types of electrodes used and some alternative methods of obtaining better connection to earth are covered in "Stage 2 Design" and so we shall not consider these again here.

So that we can be sure that our protective device will operate correctly we need to confirm the resistance of the earth electrode. We can carry out research into soil resistivity for the area and so on but until the electrode is in place we cannot confirm that it complies with the requirements. It will be useful at this point to consider the methods available for the testing of earth electrodes. Remember that these tests apply to all types of electrodes in all types of conditions and locations.

Earth electrode resistance test

We can carry out this test in a number of ways.

An a.c. supply and an ammeter and voltmeter.

We can use an a.c. supply from a double wound isolating transformer giving an SELV output which may be as low as 12 V. We connect one side of this output, with an ammeter in series, to the installation electrode. We connect the other side of the transformer output to an auxiliary electrode which we place some distance from the installation electrode. This must be installed so that the resistance area of the two electrodes does not overlap which could involve distances of 30 metres or more. A reliable figure is 10 × Installation Electrode length.

Now we install a third electrode midway between the main and auxiliary electrodes with a voltmeter connected as shown in Figure 5.6 and we can measure voltage drop between the two electrodes.

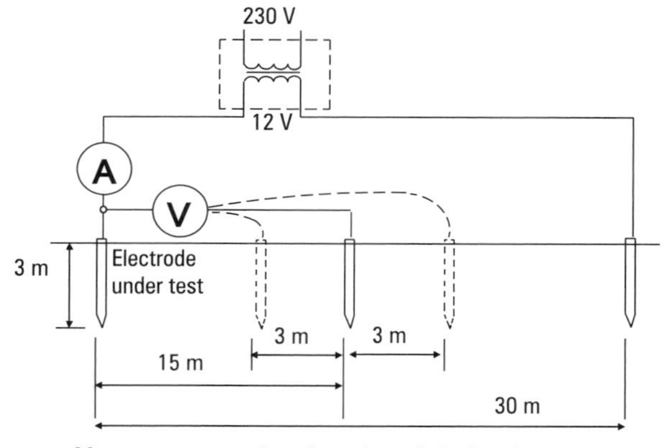

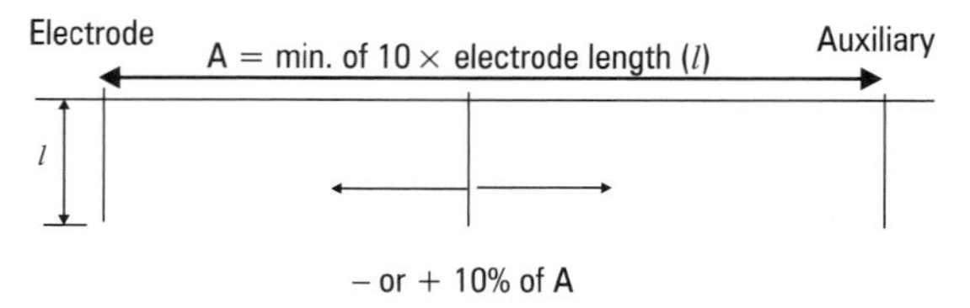

Figure 5.6 *Using an a.c. supply and an ammeter and voltmeter*

If the resistance areas of the two electrodes do not overlap, this should be the voltage drop across the resistance, to earth, of the installation electrode. In order to make sure this reading is representative we move the third electrode 3 m further from and 3 m closer to the installation electrode or 10% of the distance between the installation and the auxiliary electrode. In each case we note the results and if the readings for all three points are found to be close in value then we record the average of the three readings as the electrode resistance by dividing the average voltage by the test current.

If the results are not close in value then we must repeat the tests with the auxiliary electrode positioned further away from the installation electrode.

Remember

BS 7671 provides us with details of the types of electrodes acceptable and this includes the mention of earth rods or pipes. It is important to remember that these pipes are those driven into the ground for the sole purpose of providing a connection to earth. This does **NOT INCLUDE** the use of service pipes such as gas, oil and water which **MUST NOT** be used as earth electrodes.

An earth electrode tester

Where we have no a.c. supply available, or where it's more convenient because of the location of the electrode, we can use an Earth Ohmmeter. The instrument must produce an a.c. supply because if we use d.c. we can polarise the electrodes which will result in our readings being higher than the true values.

The procedure for carrying out the test is the same as that for the previous test, connected as shown in Figure 5.7, in this case the readings are taken directly from the instrument in ohms.

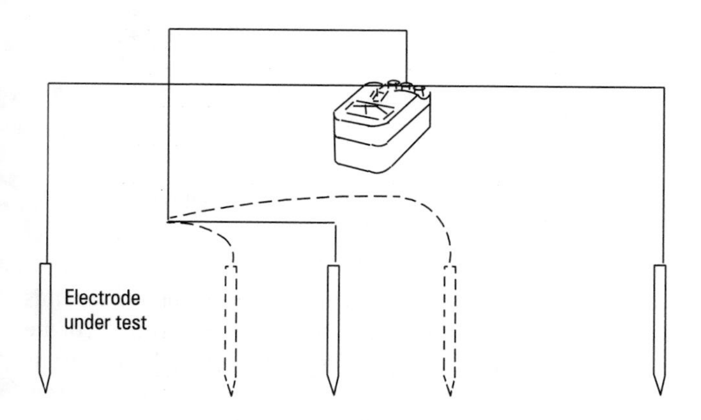

Distances for this test will be similar to those in Figure 5.6

Figure 5.7 *Earth electrode resistance test using an earth ohmmeter*

An earth fault impedance tester with electrode test facility

This is particularly useful when there is insufficient ground area to carry out either of the two previous tests. The earth fault loop path for a TT installation is via the general mass of earth back to the supply transformer. Therefore it is possible for us to measure the earth fault loop impedance with a suitable instrument connected as shown in Figure 5.8.

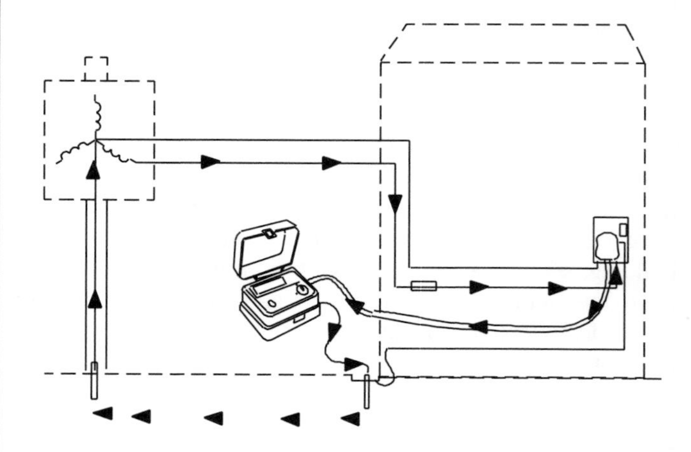

Figure 5.8 *Testing the earth electrode measuring the earth fault path which gives an indication of the earth electrode resistance.*

To simplify the drawing the consumer's and supply company's main equipment has been omitted.

If an RCD is not fitted then the electrode resistance must be low enough to cause the largest protective device installed to operate within the accepted time. A situation where the earth electrode resistance value is sufficiently low and reliable enough to ensure operation of the largest device present is extremely rare. Due to the fact that electrode resistance is subject to change and a permanent and reliable method of protection against indirect contact is required then an RCD is installed. We must then ensure that the device will operate within the requirements given in BS 7671.

Remember

The consumer's installation must be completely disconnected from the electrode whilst tests for electrode resistance are carried out.

Suitable warning notices must be posted and the users notified that the installation must not be used until it is reconnected to the electrode.

Try this

From Section 542 BS 7671 list the recognised types of earth electrode. Give typical locations or uses for 3 of the types explaining the reasons for your choice in each case.

Voltage gradients

When a fault current flows to earth, via the earth electrode, the electrode rises to a potential above that of the surrounding soil. The precise voltage will depend upon the magnitude of the fault current and the resistance of the electrode. The overall effect of this is that a voltage then exists around the electrode.

As we move away from the electrode, the resistance of the soil gradually reduces the voltage. The potential is present at the electrode until such time as the fault is cleared. During this clearance time it may cause damage to communications and telephone equipment and cables within the area. The effect of the voltage gradient in the soil may also be a real danger to persons and particularly livestock in the vicinity of the electrode.

Figure 5.9 shows how the voltage gradient from the electrode may be regarded as ripples spreading out from a pebble thrown into a pond. The greater the distance we can span the larger the potential across our body and so the risk of electric shock. Cattle and livestock are highly sensitive to shock and may be killed by voltage levels that would not create danger for humans.

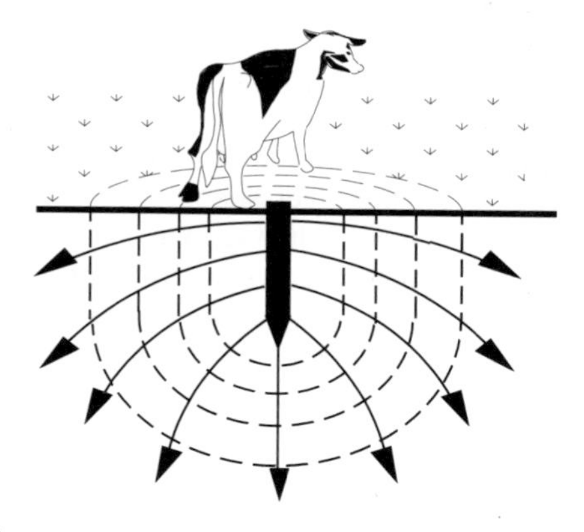

Figure 5.9

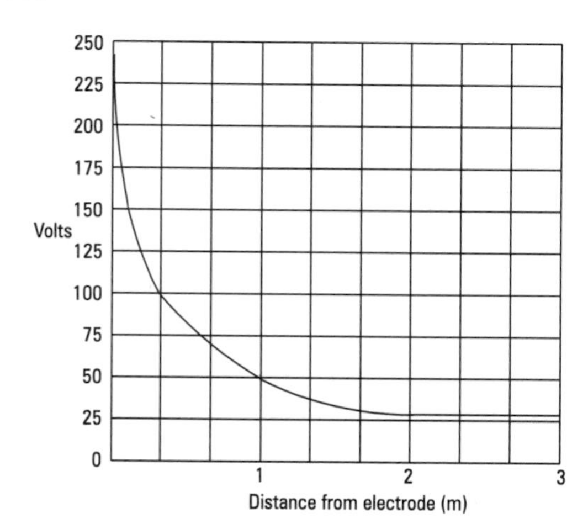

Figure 5.10 *Typical surface voltage-gradient curve around a rod electrode*

A typical voltage decay away from the electrode is shown on the simplified graph in Figure 5.10.

As livestock, by the very nature of their size, span considerable distances, this effect is a very real problem in rural and agricultural environments where electrodes are common. One way in which we can overcome this problem is to bury the top of the electrode below the surface.

Research has shown that taking the voltage gradient, measured over a span of 2 m from a 25 mm pipe electrode, at surface level, may be considerably reduced by ensuring that the whole of the electrode is buried below the surface. If we bury the top of our electrode to 300 mm and make our connection with an insulated cable, with suitable access and protection to the termination we can reduce the voltage gradient somewhere in the order of 65%. If we bury the top of the electrode to 1.1 m, we can reduce this still further to around 80%.

Bonding

We considered the use of earth monitoring devices in the previous chapter and these devices may be used on any type of earthing arrangement. If we are to use such a device it is important that all earth fault currents are carried through the main earthing conductor. In a number of instances this may be difficult to achieve due to the nature of installations and building structure.

One of the problems we face is that of "fortuitous" (accidental) earth paths which can occur at any part of the system. Let's consider the case of a lighting circuit installed in steel conduit within a building having steel supports. We can install a separate cpc within the conduit system and at some point this will connect with the metal body of our light fitting. This body will also be in contact with the steel conduit which at some point is attached to the steelwork of the building. This in turn could be in contact with the mass of earth.

In the above example we have a situation where there are three possible return paths for any fault current to earth and three possible sources of potential within the installation. The first consideration we must give is to the danger to the occupants and fabric of the building as a result of these possible differences of potential.

We overcome this by the provision of equipotential bonding which is used to connect together all the exposed and extraneous conductive parts within the building. We use main equipotential bonding to each incoming service at their points of entry into the building. We also provide a main bond to the building structure and to the lightning protection system.

We use supplementary bonding to provide additional protection in areas of increased risk, such as bathroom and shower areas, milking parlours and the like. The purpose of main and equipotential bonding is essentially to ensure that all the exposed and extraneous conductive parts remain at the same potential with respect to one another. This potential may not always be earth potential as during an earth fault the Main Earthing Terminal (MET), and hence all the exposed and extraneous conductive parts, may be considerably above earth potential.

In order for current to flow there must be a difference of potential. If by the use of main and supplementary bonding, we can maintain these parts at the same potential then there will be no current flow. The risk of electric shock through indirect contact will therefore be considerably reduced.

The provision of equipotential bonding may also reduce the possibility of multiple earth return paths. We are trying to keep all the exposed and extraneous conductive parts at the same potential. We do this by the use of low impedance connections between the exposed and extraneous conductive parts and the MET. This should encourage earth fault currents to take the path of least resistance to return to the source. If our connections are of a lower impedance than any accidental earth return path then earth fault current should flow by that route to the MET.

At this time it is worth considering the requirements for main equipotential bonding, in particular for TN-C-S systems. One reason for considering this particular aspect is that the main equipotential bonding in TN-C-S systems is also required to provide protection against circulating currents within the system. In the TN-C-S system the functions of the earthing and neutral conductors are combined within the supply conductors. This can result in some circulating currents occurring during the normal operation of the system. However should this PEN conductor become damaged or disconnected then the MET at our installation could rise to a voltage considerably higher than earth potential.

One effect of such a fault occurring on the system is that the neutral currents will seek an alternative return path, which could be via the general mass of earth. One such low impedance route to earth is via the main equipotential bonding and the other services and building structure.

This could result in a considerable circulating current passing through the main equipotential bonding conductors. This could be the neutral current for the normal load taken by the number of installations downstream of the neutral conductor break. It is likely that the neutral load current would be shared by the main equipotential bonding conductors in each installation involved. However the current carried by the main equipotential bonding conductors is likely to be considerably higher than would occur as the result of a fault on the individual installation.

It is important to remember that when this type of fault occurs the overcurrent protective devices will not respond as there is no overcurrent. This means that the diverted current may flow for some considerable time period. This is a very different scenario to the installation fault which should result in the circuit involved being disconnected within the prescribed time period from BS 7671.

As a result the cross sectional area for the main equipotential bonding conductors in installations forming part of a TN-C-S system are dependant upon the cross sectional area of the incoming supply neutral conductor. BS 7671 provides guidance in Section 547 on the selection of main equipotential bonding conductors. This requirement ensures that the main equipotential bonding conductors are sized to take account of this possibility and to carry such currents without risk to the installation, the user or the fabric of buildings and equipment.

It is worth noting that some Public Electricity Suppliers may require larger conductors that those in BS 7671 and so it is always worthwhile establishing their specific requirements before installing the main equipotential bonding conductors.

Exercises

1. Describe, with the aid of diagrams, one method of carrying out an earth electrode resistance test. This should NOT be using an earth fault loop impedance tester.

2. Explain the effect of voltage gradients surrounding an earth electrode and list actions that can be taken to minimise the risks resulting from these gradients.

3. With regard to bonding
   (a) explain the reasons for main equipotential bonding
   (b) explain the reasons for supplementary equipotential bonding
   (c) list the locations where supplementary equipotential bonding is necessary giving your reasons why this is so.

4. Explain the purpose of earth monitoring systems and the effect of fortuitous earth paths on such a system.

6

Metering and Tariffs

Revision questions:

1. Explain briefly what requirements the Electricity Supply Regulations 1988 make upon the Public Electricity Supplier in respect of earthing:
   (a) the system and
   (b) the consumer's installation.

2. List the three main types of supply systems available from the Public Electricity Suppliers.

3. Sketch the earth fault loop impedance path for a TN-C-S system labelling the component parts.

4. What is the purpose of carrying out an earth electrode test?

5. List three methods by which earth electrode resistance can be measured.

6. When earth electrodes are installed in agricultural environments what particular feature must be given careful consideration regarding the location of the electrode?

7. Identify the two types of "bonding" which would be carried out in connection with the electrical installation in a consumer's premises and an indication as to where these would be applied.

On completion of this chapter you should be able to:

- describe meter arrangements for domestic and industrial installations
- describe the methods of metering and monitoring supplies
- explain the types of tariffs available and their advantages
- draw the circuit diagrams for typical meter connections
- calculate costs of electricity using current tariff rates

In this chapter we are going to consider the requirements for the metering and monitoring of supplies both by the supply authority and the consumer. We shall also look at the tariffs available and their advantages and limitations.

We shall first look at the supply companies' requirements for metering and costing for electricity used by the consumer. The cost of producing electricity can be split into two main areas, the standing and running charges.

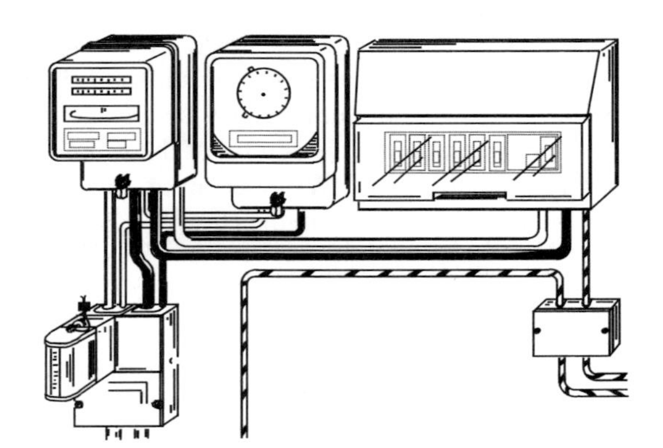

Figure 6.1 *Typical domestic metering arrangement for TT system*

Standing charges

These cover the cost of plant and equipment purchase for the production of electricity, the cost of staff, both technical and administrative. It also covers the rent and rates for buildings and land including that paid for the siting of pylons and substations. All these are budgeted for by the supplier and incorporated in the standing charge portion of the tariff.

Running charges

This includes the cost of fuel and disposal of waste, replacement of plant and maintenance costs. These items are subject to short term changes and are less easy to forecast. Also included in the running cost is the profit, which is necessary to provide for future development and expansion.

Electricity Suppliers purchase electricity in bulk from the generating companies via the National Grid network, they then sell this electricity on to their customers. The cost of transmitting the electricity from the generating stations, through the National Grid network and then distribution to the customer via the local distribution system, is also borne by the supplier.

The suppliers publish their prices, read the meters, send out the bills and deal with the customer on a day-to-day basis. Any organisation can enter this market provided they can obtain a licence to do so. The issue of the licence is evidence that the company or organisation concerned is considered competent to carry out the business.

Maximum demand

The main purpose of the maximum demand tariff, as far as the Public Electricity Supplier is concerned, is to encourage the consumer to spread the load over a period of time and not to load the system to excess. This is a two-part tariff, the first part being a fixed charge based on the maximum kVA demand. A further charge is made for the units as indicated on the kWh Meters. The better the power factor of the installation the less difference there will be between the kWh meter reading and the actual power consumed.

A typical arrangement for a maximum demand meter is shown in Figure 6.2. The switching device may be a standard time switch or a tele-switch under the control of the supplier. This meter will register the total kWh over an arranged period, divided by the number of hours in the period.

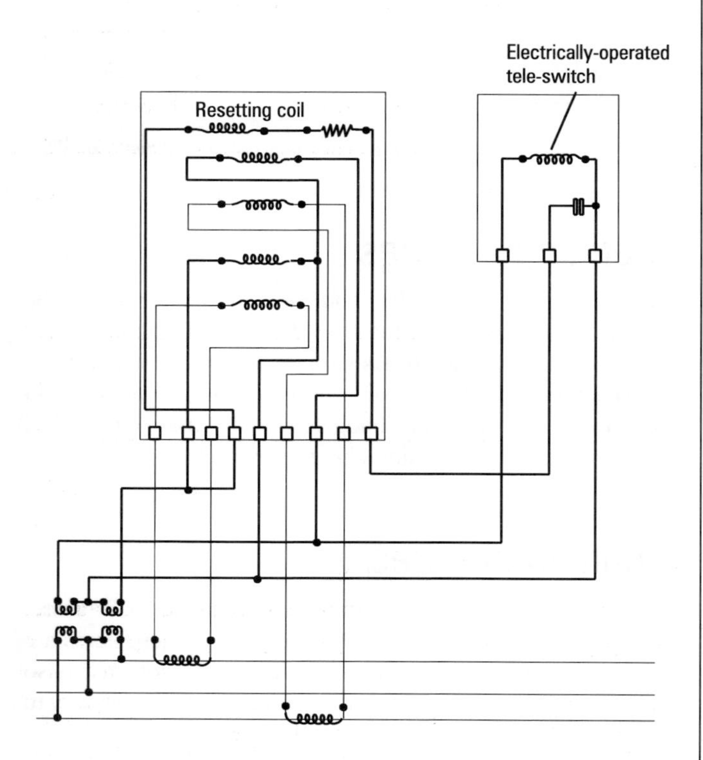

Figure 6.2 Maximum demand meter internal connections

The maximum demand is the average demand over a short period of time, usually 30 to 45 minutes. The recording device for this is best considered as an analogue measure as shown in Figure 6.3.

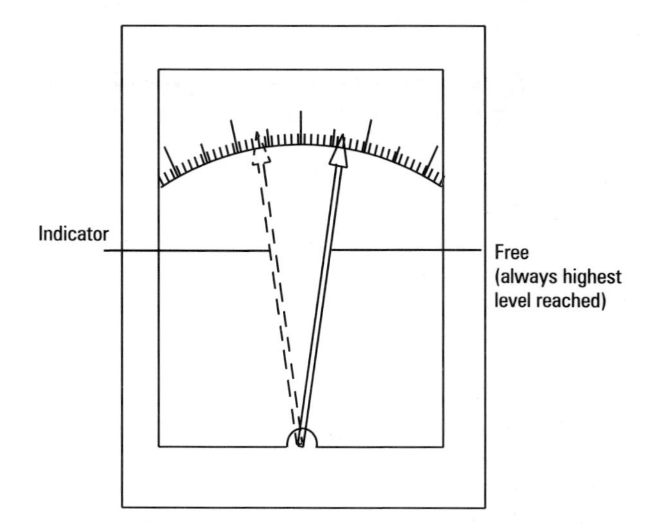

Figure 6.3 Analogue measure for maximum demand

There are two indicators one of which registers the maximum demand during the recording period and returns to zero at the end. The other is a free indicator which is pushed round by the indicator. As this second pointer has no return mechanism, it will remain at the furthest position it reaches. This means that if at the next recording period the demand is higher the free indicator will be pushed further round the scale. If the demand is lower the free indicator will remain in the last highest position. In this way the free indicator will always show the maximum demand occurring over the entire period.

The consumer pays a high cost for the maximum demand, so there is an incentive to spread the load to ensure that the maximum demand is kept to a minimum. If a large manufacturing concern starts up every day at 07.30 and all the machines and lights are switched on at the same time, the resulting starting current may be more than six times the normal running level. By staggering the start up time of groups of machines and equipment this maximum demand may be kept to a minimum, offering a considerable saving to the company.

Cheap rate tariff

These were originally introduced in the late 1960s to encourage consumers to use electricity during the "off peak" periods. These are the times when the generation of electricity must be continued and demand is at its lowest, the cost and time involved in shutdown of a conventional fossil fuel or nuclear generator makes stopping them impractical. The period was generally between around midnight and seven in the morning. To encourage the use of electricity during these hours the promotion of "off peak" storage heating was increased and the design of the heaters improved making them smaller and lighter. The development of these heaters has continued and the latest are slim and lightweight, with the facility to provide both off and on peak heating. This is generally through the provision of either convection/radiation or fan assisted heating elements for the on peak use. Such heaters will require two circuits, one for the off peak, the other to provide the on peak supply.

To further encourage their use the supply companies offer a dual rate tariff. This provides a cheap night time rate and a daytime rate set slightly above the single rate tariff. The electricity consumed during the day was metered and charged at a higher rate than the night-time units which are usually charged at less than half the rate of the daytime units. This process was carried out by a conventional watt-hour meter with two sets of recording counters.

This "dual rate" meter incorporated a small, solenoid operated, change-over from one counter to the other. This changeover was originally controlled by a timeswitch supplied by, and under the control of, the Public Electricity Supplier. A typical example is shown in Figure 6.4. The principal advantage of this system is that at the changeover from the day to night rate **all** the on peak circuits are automatically transferred to the cheaper rate. Only those circuits intended for off peak use, such as the storage radiators, need to be separately controlled as these will not be operational during the on peak times.

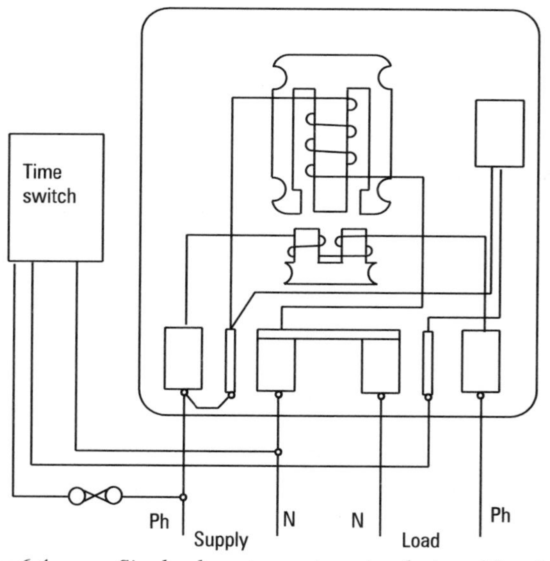

Figure 6.4 *Single phase two-rate meter designed for cheap off-peak tariffs*

It is possible to provide for "dual control" of certain equipment, such as water heaters, with dual elements. Here two elements are provided, a lower element which heats the entire content of the hot water cylinder during the cheap rate. A high level, on peak, element is also installed to maintain the top portion of the tank at the desired temperature, should the demand for hot water exceed the off peak provision.

Developments in technology have allowed this type of tariff to evolve considerably to a highly sophisticated "total control" tariff offered by some of the supply companies. This system allows for the use of both on peak and off peak supplies with the added advantage that the supply company can provide a boost to the off peak at any time when their demand falls. This is subject to the consumer receiving the agreed minimum period of cheap electricity in a 24-hour period.

This has been made possible by the use of the tele-switch, which can be controlled by the supply authority on the transmission of a signal, in place of the fixed period time switch. This also allows the supplier to switch consumers on a staggered basis and so prevent the sudden increase in load which was experienced with the old time switch system.

Typical multi-rate tariff

Modern meter technology has allowed suppliers to offer more varied rate tariffs than before with a minimum of control equipment. A typical tariff will now offer three separate component rates, a single rate for "off peak" heating, and a two-part rate for daytime use. The application of these rates can best be shown in pictorial form and Figures 6.5 and 6.6 show typical times and cost per unit for this type of tariff.

Figure 6.5 shows that, as a result of the advances in tele-switch technology, the consumer is offered different rates for electricity at different times of the day and night with an "all day" lower rate for weekends and bank holidays. These times will be reasonably consistent and will be seasonally adjusted so that the switching will occur at the same time both summer and winter, i.e. if switching is at 07.00 it will occur at that time for both BST and GMT.

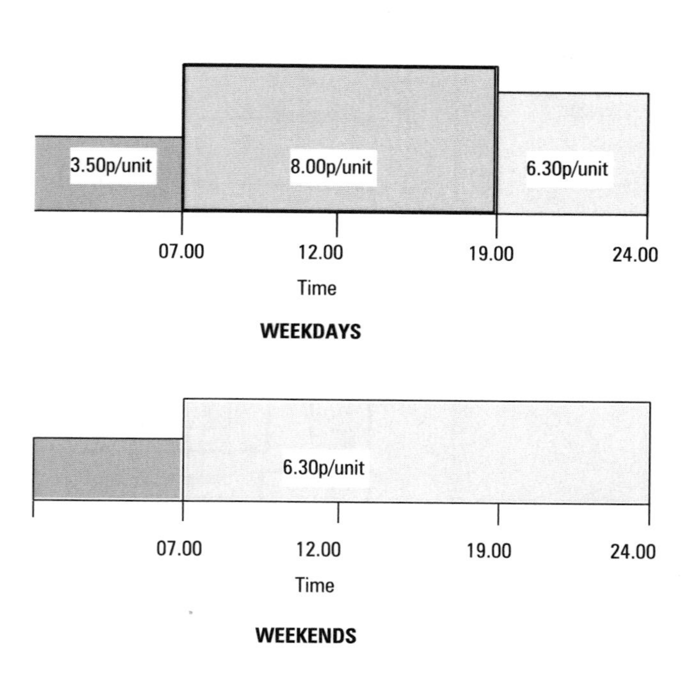

Figure 6.5 *Unrestricted supply for normal circuitry*

Figure 6.6 shows a typical arrangement for the storage heating load. The supply company undertakes to provide a total of 8 hours of electricity at this rate. Of these six hours will be between 22.00 and 08.00 and a further two hours between 12.00 and 17.00 at the discretion of the supply company.

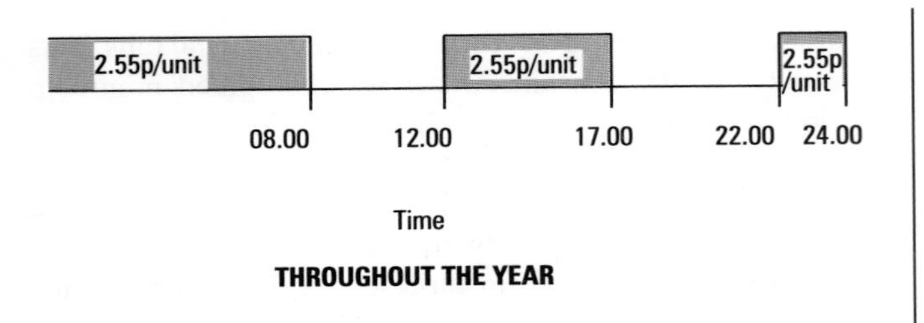

Figure 6.6 *Restricted supply for Off Peak heating load*

The typical arrangement for the connection of this type of installation is shown in Figure 6.7. which, as we can see, is a new improved version of the two rate tariff.

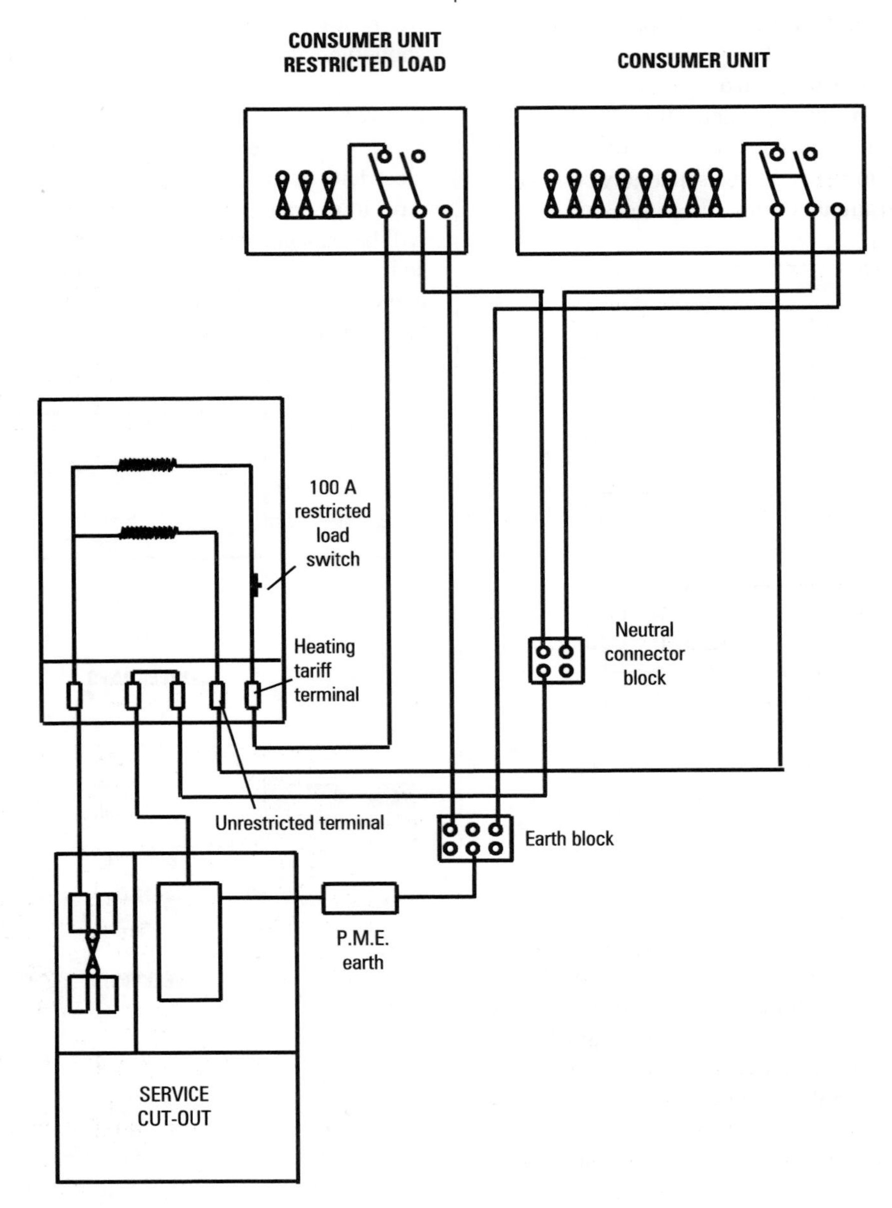

Figure 6.7 *Typical multi-rate metering*

Standard tariff

A standard tariff is still available for consumers who wish to use it, where the cost charged is still based on a fixed charge and a cost per unit for electricity consumed.The variety, but not flexibility, of tariffs now offered has considerably reduced and suppliers tend to standardise on the types we have considered here. It is normal for large consumers to discuss wit ier the most beneficial tariff for the type of use they will have. In this way the supplier is aware of the nature and period of probable demand and can make the tariff attractive to the consumer. A built in control factor to encourage the maintenance of a high power factor and the restriction of maximum demand may also be incorporated.

Power factor

We are aware of the effects of power factor and power factor improvement on energy consumption. The supply companies regard a low power factor, which tend to exist on the majority of industrial and some commercial installations, as increasing their costs with no return for the additional outlay.

We know that the effect of the reactive components causing a low power factor is to increase current drawn from the supply. This will require an increase in the size of cables and equipment to supply this demand. At the same time the power drawn from the supply in kWh, and the resulting revenue, does not increase. The cost per kWh unit does not, therefore, take into account the poor power factor and the resultant increased cost to the suppliers.

Modern equipment and installations are designed to ensure that a high power factor is maintained and this may be achieved in a number of ways. The use of power factor correction capacitors fitted to machines and equipment for example. This ensures that the power factor correction is connected in the circuit only when the equipment is in use. Another option is the use of high frequency fluorescent lighting units which also provide an improvement in power factor over the conventional fluorescent lights.

Connection of energy meters

Whole current measurement

The use of whole current metering is normally restricted to smaller installations, up to say 100 A. With this type of metering the whole of the current drawn from the supply passes through the meter as shown in Figure 6.8. Metering in this way is restricted because of the capacity of the meter and the resulting physical size. We could have a 600 A meter of this type, however the physical size of the meter would make it impractical. A further problem with this type of metering is that a failure in the meter may result in the loss of supply to the installation.

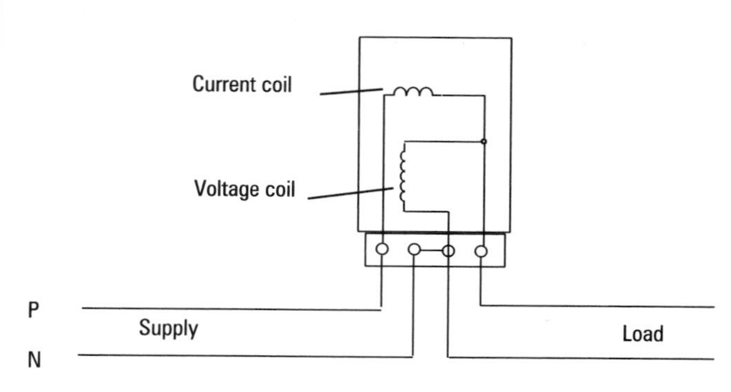

Figure 6.8 *Whole current kWh meter*

This type of meter may be used to measure single and poly phase installations. Some three phase installations are metered as shown in Figure 6.9 but it is more common to have a single, poly phase meter.

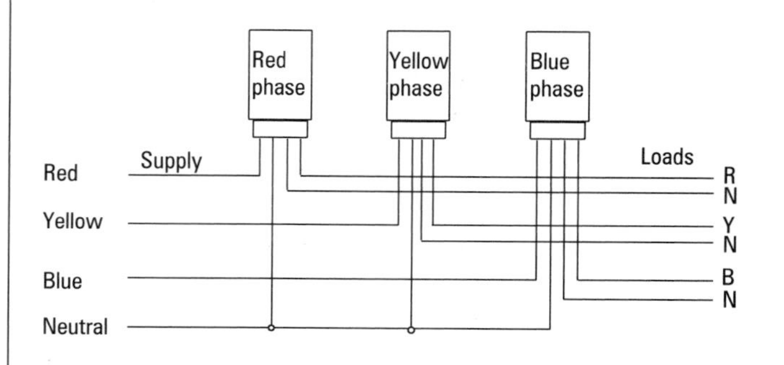

Figure 6.9

CT and VT meters

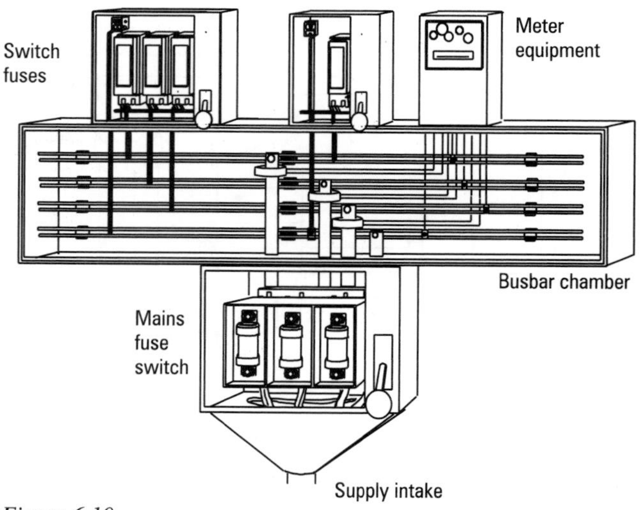

Figure 6.10

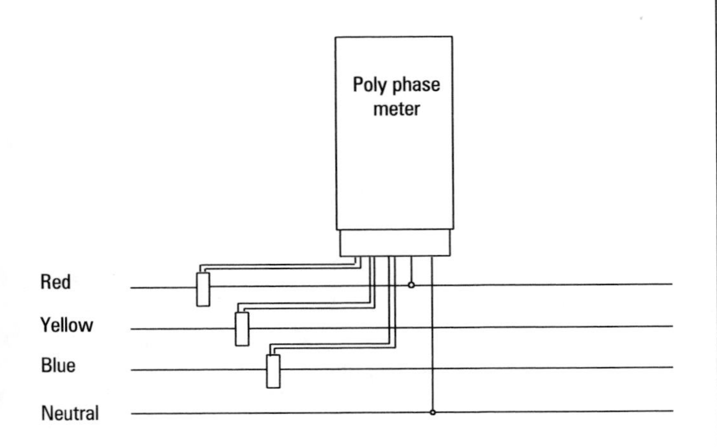

Figure 6.11 *Poly phase CT connection*

It is increasingly common practice to measure the electricity consumption of consumers via transformers as shown in Figures 6.10 and 6.11. This has a number of advantages in so much as

- the capacity of the meter can be considerably smaller
- the physical size of the meter can be reduced
- any failure of, or maintenance to, the metering equipment does not affect the continuity of supply to the consumer
- increase in demand from a consumer can be more easily accommodated

all of which makes this type of metering popular with the supplier and consumer. As we can see, this type of arrangement does work very well with the metering requirements for TP & N installations.

Summation metering

The purpose of summation metering is to allow a single instrument to record the total consumption for a poly phase installation. The basic circuit diagram for this configuration is shown in Figure 6.12.

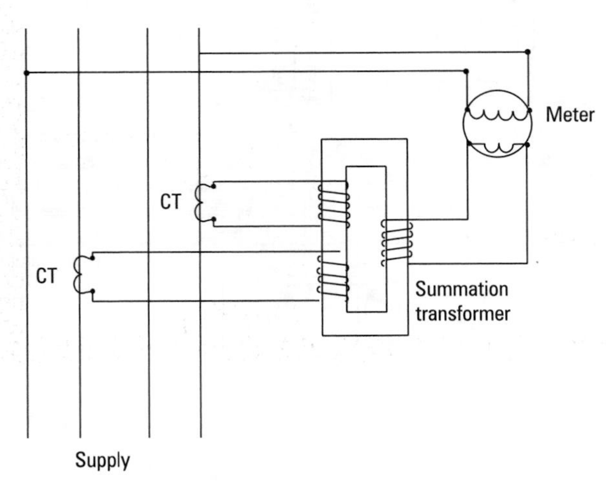

Figure 6.12 *Simple summation meter*

The principle of operation is that current transformers are used to connect the main intake supply conductors to the current coils on the instrument. The instrument has its current coils mounted as the primary of a summation transformer. The current produced in the secondary of the transformer is the phasor sum of the currents flowing in the primary windings.

The use of this type of arrangement allows a small single reading meter to be used to record the total consumption for a large industrial consumer. The actual size of the meter itself need be no larger than a standard domestic meter, although the ancillary equipment will usually require a somewhat larger enclosure.

Metering requirements

When the supplier is to carry out metering using any of the above methods some consultation is necessary with the consumer or the designer of the installation. The supplier will have a need for certain features to be included to allow the metering to be carried out.

The typical points to be considered are

- physical location of metering equipment
- means of connection to the consumer's installation, e.g. where the CT's are to be mounted
- space required for metering equipment
- security of and access to metering equipment

Both parties need to be aware of the supplier's needs and the necessary arrangements will need to be made to agree the final locations before the work is undertaken and switchgear and so on ordered.

Tariffs

We have already considered the types of tariffs offered by the suppliers for the consumption of electricity. Let's now take a look at the application of these tariffs.

Domestic "Economy Seven" tariff

This involves the consumer in payment for fixed charges, electricity used during normal times and electricity used at off peak times. It is usually referred to as the Economy Seven tariff because the supplier undertakes to provide at least seven hours during every 24 hours at a reduced "economy" rate.

The electricity bill is made of the three components and a typical domestic bill is shown in Figure 6.12 and we can see the amounts detailed as the number of units used and the price per unit added to which we have a standing charge. The total amount is then subject to VAT and an overall cost is produced.

Many of the old tariffs have now been removed and the process somewhat simplified in recent years. Whilst smaller consumers have less choice, the changes in technology have allowed the supplier to offer more attractive rates as a result of much closer and more reliable control.

....ELECTRICITY

Account Number
Date (Tax Point)
Business Use
0%
Reading Date

Electricity Bill

Meter Readings		Units	Unit Price	VAT	Amount
Present	Previous	Used	(pence)	Code	£
E 07683	C 07473	210	7.88	1	16.55
E 40721	C 40431	290	2.38	1	6.90
Standing charge			TO	1	12.20
TOTAL CHARGES		(EXCLUDING VAT)			35.65
VAT1	£35.65 @	5% DOMESTIC			1.78
			TOTAL		37.43

VAT charge this bill 1.78

PAYMENT DUE
£37.43

E = Estimated reading. If you are not happy with this reading, please phone us now.
C = Your own reading.

Figure 6.13 Typical domestic two-rate electricity bill

Try this

A consumer on a two-rate economy tariff has the following consumption over a thirteen-week period:

- 497 units at daytime rate (7.6 p.p.u.)
- 2360 units at economy rate (2.8 p.p.u.)
- standing charge £14.50

During the course of the year there are two thirteen-week periods at this level of consumption and two where the daytime use is reduced by 5% and the economy use is reduced by 60%.

Calculate the annual electricity charges made to the consumer on this tariff and the equivalent cost if the units were charged on a single tariff rate of 6.5 p.p.u.

Maximum demand

When a consumer requests a supply on this tariff one of the details required by the supplier is the estimate of the probable highest demand. This is likely to be included in an additional charge based on the "availability" of electricity to that consumer. This is the standing charge to this consumer and is intended to cover the cost of capital outlay and equipment required to make the required supply available.

It is common for the supplier to set a period of around 5 years from the time of connection before this figure can be reduced. As the standing charge will be based upon this level for the 5-year period, it pays to be accurate with this assessment. If the actual demand exceeds the estimated figure, at any time during the period, then a revised level is set. From that time onwards the costs are based upon the revised level of maximum demand.

The charges to these consumers are usually made on a monthly basis and are made up of Standing Charge, Availability Charge and Maximum Demand charges. The standing charge is a flat fee but the others are calculated on a "per kVA" and "per kW" basis.

Other factors such as the time of day and the time of year may also affect the charges made to the consumer. It is therefore a good idea to consult closely with the supplier to establish the best tariff for the consumer's needs. Consumers may now select which supplier they wish to use no matter which area of the country the premises is located. This makes the investigation of the available tariffs even more important.

We discussed earlier the possibility of the purchase of energy at low or high voltage. The table shown in Figure 6.14 shows the comparable cost and saving between the two options. Two loads have been used and losses in the transformer, which will be at the cost of the consumer, have been allowed in the HV calculations. The maximum demand has been set at 370 kVA.

We can see that the consumer will need to be using considerable amounts of electricity to recover the cost of installing HV switchgear and transformers etc. It is usually the very large consumer, and the consumer who is looking for stability and control of the supply, that will take this option.

	Supply at Low Voltage		Supply at High Voltage	
Average load, kVA, in factory 3800 hr/year	340	200	340	200
Standing charge per year Availability charge per year Max. demand charge per year (average of monthly charges)	78.05 × 12 0.94 × 370 × 12 4.16 × 12 × 370	936.60 4173.60 18500.00	255.93 × 12 0.89 × 370 × 12 3.79 × 12 × 370	3071.16 3951.60 16835.00
Total fixed charges	23610.20		23857.76	
Unit charges at 340 kVA average load in factory; 1,292,000 kVA	63049.60		57752.40 (347 metered at HV)	
Unit charges, at 200 kW average load in factory; 760,000 kVA		37088.00		33972.00 (207 metered at HV)
(plus fixed charges)	23610.20	23610.20	23857.76	23857.76
Total charges £ per year	86,659.80	60.698.20	81.610.16	57,829.76
Cost per unit entering factory in pence	6.7p	7.98p	6.3p	7.6p
Saving in purchased power by using HV tariff			5049.44	2868.44

Figure 6.14 Typical comparison for purchase of electricity at LV and HV (VAT not included)

Try this

Take the figures used in Figure 6.14 and apply the current tariff from your local electricity supplier to establish the level of saving your local industrial consumer would expect to make.

Exercises

1. Explain the advantages of metering industrial consumers via CTs and VTs.

2. Draw the circuit diagram for a typical multi-rate domestic intake identifying each component.

3. Explain the principle of maximum demand tariffs and the steps that can be taken by the consumer to minimise costs.

4. A consumer on a two-rate tariff has the following use each 13-week period throughout the year:

   35 units at standard rate

   1147 units at off peak rate

   Using your local PES current tariff calculate the annual cost of energy based on this consumption. Based upon the result what recommendation could you make to the consumer to help reduce this costing.

End Questions

1. Draw the circuit diagram for a ring distribution system supplying 3 substations. Show the necessary switching and interconnections to allow the system to be fully maintained without interruption of the supply.

2. (a) Describe two methods of cooling transformers giving the advantages and disadvantages of each.
   (b) A factory extension necessitates the installation of an additional substation. This is to be located outside of the factory building. From the two types used for (a) above select the most suitable for this application giving details of the type of substation construction you would recommend and reasons for your choice and requirements.
   (c) Draw a simple layout of the substation for (b) above labelling each component.

3. (a) Describe, with the aid of sketches, the method of arc control for the following devices
   i) oil-filled circuit breaker
   ii) air blast circuit breaker
   iii) vacuum circuit breaker
   iv) BS EN 60898-1 type fuse
   (b) With the aid of diagrams explain how CT's may be used in conjunction with circuit breakers to provide earth leakage protection for distribution networks.

4. (a) Describe, with the aid of diagrams, the method of earthing for TT, TN-S and TN-C-S systems and define the earth fault loop in each case.
   (b) Explain the effect of voltage gradients on livestock and detail methods by which these effects can be reduced.

5. (a) Describe three methods of installing a distribution system within two storey commercial premises giving the advantages and disadvantages of each.
   (b) Calculate the maximum demand for the above installation, after application of diversity in accordance with Guidance Note 1, if it comprises
   40 × 6 A lighting circuits
   15 × 32 A ring circuits
   2 × 8 kW storage water heaters single phase
   4 × 9 kW instantaneous showers single phase
   14 × 16 A motor circuits
   and the supply is 400/230 V.
   (c) Assuming that the above load, after diversity, operates for 6 hours a day, 5 days a week, with all water heating operating for 10% of this time, and the power factor is unity calculate
   i) the total consumption over 13 weeks
   ii) the cost of the units consumed on a single tariff at 6.8p per unit.

Answers

These answers are given for guidance and are not necessarily the only possible solutions.

Chapter 1

p.5 Try this: (1) As detailed in the Electricity Supply Regulations 1988, Regulation 34, paragraphs 2 to 5
(2) This requirement is identified in Regulation 36 of the Electricity Supply Regulations 1988

p.6 Exercises: (2) The following regulations apply to the nature of the supply authorities' requirements and obligations:
Regulation 2 Application,
Regulation 3 Interpretation
Regulation 5 General requirements for earthing
Regulation 7 Protective multiple earthing
Regulation 8 Earthing metalwork
These next regulations may affect the design, installation and construction of the consumer's installation:
Regulation 20 High voltage additional requirements
Regulation 25 Supplier's works on consumers' premises
Regulation 27 General conditions as to consumers
Regulation 28 Discontinuation of supply
Regulation 30 Declaration (including Amendment 2 1994)
Regulation 31 Information on request
Regulation 38 Works in breach
Schedule 1 Safety sign.
There may be other regulations applicable particularly if alternative supplies are being considered. A knowledge of the requirements of the Electricity Supply regulations is essential when carrying out the design of such installations. (3) The details of the actions required are set out in Regulation 38 and any consumer who fails to comply with this regulation is guilty of an offence under Section 16 of the Energy Act 1983 (Regulation 39 refers).

Chapter 2

p.7 Revision questions: (1) Where the consumer has their own generation equipment which is connected to the supplier's equipment, albeit through a switching arrangement. (2) (a) the number and rotation of phases (b) the frequency (c) the voltage (d) maximum variation in frequency ± tolerances (e) maximum variation in voltage ± tolerances (f) maximum prospective short circuit current at the supply terminals (g) maximum earth fault loop impedance of the earth fault path outside the consumer's installation (h) type and rating of the supplier's fuse or switching device nearest the supply terminals (3) The Public Electricity Supplier's particular requirements for PME, protective multiple earth, (TN-C-S) systems

p.10 Try this: Switching sequence is as follows:
At Substation 4
Close breaker 4c then close breaker 4f and finally close link switch M4
At Substation 3
Close breaker 3c then close breaker 3f and close link switch M3
Now all six transformers are supplying the load and we can isolate the transformers 3A and 3B.
At Substation 3
Open 3d then open 3e now open 3a and then open 3b.
Transformers 3A and 3B are now isolated from the system and the load is being supplied via transformers 3C and 4A, 4B and 4C. There are 4 transformers supplying the total load from substations 3 and 4 so the system has not been overloaded and there has been no interruption of supply.

p.13 Try this: (1) 12.077 A; (2) 45.1 A

p.18 Try this: 21.3 A

p.20 Try this: (1) (a) increases resistance in cable – less current can flow; (b) cable size needs increasing to reduce heat generated in cable. (2) All cables will produce heat – less heat dissipated – less current can flow. (3) 12.42 A

p.23 Try this: 6 mm^2

p.23 Try this: (1) ambient temperature, grouping, thermal insulation, semi-enclosed fuse used. (2) Max. 8 mV/A/m. (3) 4 mm^2 at 10 mV/A/m. (4) max. load 20.689 A

p.27 Try this: (1) 24.20; (2) 14.82; (3) 10.49

p.30 Try this: (1) 454.58 A; (2) 36.99 mm^2; (3) (a) 980 A, (b) 0.6 s

p.32 Try this: (1) Typically:
– determination of load by use of actual equipment to b e installed/used, for example domestic or warehouse for storage
– determination of load per square metre based upon the building use, for example spec. office block or small hotel
(2) I_b – the design current drawn from the supply under normal conditions. (3) The number and types of conductor, the type of earthing arrangement available, the nominal voltage, the nature and frequency of the current, the prospective short circuit current at the origin of the installation, the type and rating of the overcurrent protective device, the suitability of the supply, the earth fault loop impedance. (4) 16.75 A. (5) 0.71

p.33 Exercises: (1) (a) Symmetrical 57.16 MVA; (b) 25.46 kA. (2) The solution should include from the following: (a) Advantages: less complex system, lower initial installation cost, lower initial plant cost, lower material and maintenance cost and low cost of spares to be held
Disadvantages: reliability of supply, isolation for

maintenance , lack of flexibility and reduced security of supply; (b) advantages: reliability, ease of maintenance, flexibility and security of supply; disadvantages: cost of installation, cost of maintenance and management of a more complex system

p.34 (3)

Lights	140.4 A
Ring	208.0 A
Water heaters	92.3 A
Motors	278.4 A
Maximum demand	719.1 A
+10%	791.01 A
over 3 phase	=264 A
∴ 300 A 3 phase	

Chapter 3

p.35 Revision questions: (1) The answer may include one of the following:
Ring main, advantages: reliability, ease of maintenance, flexibility, and security of supply, disadvantages: cost of installation, cost of maintenance and complexity
Radial, advantages: cost of installation, low cost of spares to be held, disadvantages: reliability of supply, isolation for maintenance, lack of flexibility and reduced security of supply: (2) The answer may include: lower rates from the supplier, greater flexibility, reduced energy losses in the distribution cables, greater reliability of supply and distribution system under the user's control. (3) See p.11. (4) Unbalanced loads are reflected back through the system and cause imbalance of the network. The supply needs to be capable of providing the highest demand and this load then must be reflected across all phases. The effect of unbalanced loads can cause additional stress on equipment and inefficiency in operation.

p.42 Try this: see table on p.41

p.43 (1) See p.40

p.46 Exercises: (3) Details for these precautions can be found on p.41 (cable routes), p.42 (single core power cables) and p.43 (cable ducts and sealing of cable ducts).

Chapter 4

p.47 Revision questions: (1) the material from which the conductor is made, the cross sectional area of the conductor and the length of the conductor. (2) the answer could include: copper, advantages: higher conductivity, smaller c.s.a., smaller diameter cables, cheaper jointing methods, and no electrolytic action with brass terminations, disadvantages: cost and weight Aluminium, advantages: cheapness, lightweight and easier handling, disadvantages: larger c.s.a., larger cable diameter, larger bending radius, electrolytic reaction with brass and more space required to accommodate terminations. (3) Single core steel armoured cables should not be used for a.c. supplies due to the inductive effect. Non-ferrous or unarmoured cables with suitable mechanical protection should be used. However, these may also cause problems with EMC emissions and care should be used when using such cables with particular regard to siting and screening. (4) A d.c. voltage of the appropriate value (refer to the British Standard) between the conductors and between the conductors and the metallic sheath or armour. The test to be carried out once the jointing and termination is complete but before connection to the system. The voltage is to be applied gradually and must be maintained for a period of 15 minutes without breakdown of the insulation.

p.49 Try this: The solution could include either a Silicone or Dry class C. The most appropriate would normally be the Silicone, as the Dry class C is more susceptible to the climatic conditions which may affect roof mounted plant. Oil filled is definitely not an option. See p.48.

p.50 Try this: Reduction in contact pressure causes increased resistance, rise in temperature due to $I^2 R$; losses produce thermionic emission. Further opening of the contacts produces voltage across gap which has higher resistance and this increases ionisation. This produces sufficient electrons to continue the current flow across the gap and sustain the arc. The process produces considerable amounts of energy.
Contributory factors include: current flowing in the circuit, voltage, speed of operation, inductive loads and power factor.

p.53 Try this: see pp.51–53

p.56 Exercises: (2) see p.50; (4) see pp.54 and 55

Chapter 5

p.57 Revision questions (1) The Electricity at Work Regulations 1988 (2) Oil filled, Silicone and Dry class C (3) air blast, oil, gas and vacuum (4) The above circuit breakers plus moulded case circuit breakers, BS EN 60269-1 (high breaking capacity) fuses, BS 1361 (cartridge) fuses and BS 3036 (semi-enclosed) fuses. Arc control methods are detailed on pp. 51–53. (5) descriptions are given on pp.34 and 35.

p.60 Try this: (2) (a) There are two principal options available. You isolate the installation from the supply; however this may cause some considerable hardship to the consumer. A better option would be to install an RCD to protect the users of the installation as a temporary measure. The external EFLI is outside of our control, disconnection within the time required by BS 7671 cannot be achieved and the installation is unsafe, situation regarded as similar to TT system for both effect and cure. (b) Notify the Public Electricity Supplier of the situation and request that a permanent reliable earth is provided to the installation. It may transpire that if this cannot be achieved then the RCD will need to be a permanent part of the consumer's installation. Z_e is under the control of the Public Electricity Supplier and we cannot influence or control this. The PES has to be advised in writing in order that they can take the appropriate action in remedy the situation or advise that this cannot be done and alternative arrangements must be made.

p.65 Exercises: (1) see p.62; (2) see pp 63 and 64

p.66 Exercises: (3) see p.64; (4) see p.64 and Chapter 4

Chapter 6

p.67 Revision questions: (1) (a) the supplier must ensure that the system is connected to earth; (b) the supplier is not obliged to make an earthing facility available to the consumer. (2) TT, TN-S and TN-C-S.

(3)

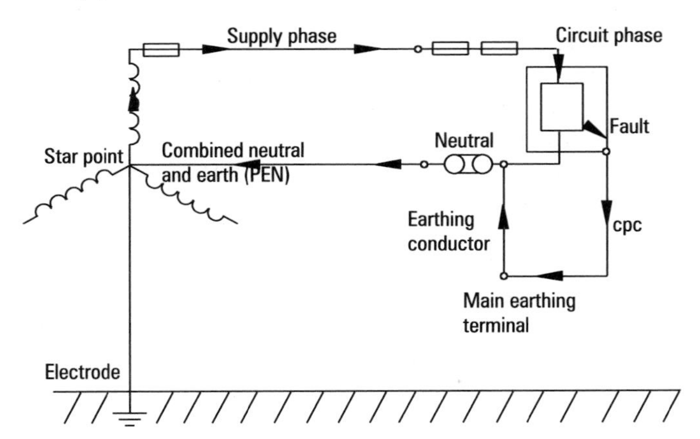

(4) to ensure that the electrode provides a sufficiently reliable connection to the mass of earth. Also to confirm that the electrode resistance is sufficiently low to ensure operation of the protective device in the event of a fault to earth. (5) An a.c. supply and an ammeter and voltmeter, an earth electrode tester (earth ohmmeter) and an earth fault loop impedance test instrument. (6) The need to ensure that the risk of livestock being affected by voltage gradients around the electrode is minimal (7) Main equipotential bonding: to incoming services and steel structures, supplementary bonding: in areas of increased risk such as bathrooms.

p.74 Try this: two rate tariff cost £421.55, single tariff cost £662.59

p.77 Exercises: (1) see p.52; (3) see p.55

End Questions

p.58 (5) (b) 783.4 A; (c) (i) 56756.7 units, (ii) £3859.46